Architectural Engineering and Design Management

ENVIRONMENT

Boussabaine

...rthscan
...ing for a sustainable future

Contents

Editorial

Halim Boussabaine

Theories, implications and applications of complexity and complex adaptive systems have grown enormously since the mid-20th century. Emerging out of the natural sciences and increasingly spilling over into the social sciences and arts, they offer a unique interdisciplinary framework for linking the often separate worlds of natural, social and artistic studies, going beyond the unnecessarily rigid boundaries of individual disciplines and exploring the untapped potential of intellectual crossovers and multidisciplinary interaction. As the Nobel Prize winner in chemistry, Ilya Prigogine, wrote, humanity is:

> *...at the beginning of a new scientific era. We are observing the birth of a science that is no longer limited to idealized and simplified situations but reflects the complexity of the real world.*

Complexity science provides more than just metaphors for comparing management situations with those found in the natural sciences; its use can produce substantial benefits. It is suggested that by using complexity tools, organizations like Citibank Credit Risk uncovered $200m in hidden expenses, Proctor and Gamble saved 22% on their distribution expenses and Dupont saved $500m annually in manufacturing expenses (Dunn, 2003).

As society's dependence on the finite resources of the built environment increases, the magnitude and complexity of problems associated with the use and management of built environment assets become more apparent. The issue of complexity in the built environment has grown significantly, providing several key application areas, including sustainable development, infrastructure, supply chain, networking, energy utilization, project management, etc. By its very nature, the built environment is a complex multi-disciplinary field that requires integrating expertise and input from various jurisdictions. Complexity theory is poised to propel research into the built environment beyond the traditional approaches. The study of complex systems has grown along many seemingly independent strands and, as with nearly anything that is new, it grew in a disorganized way. Complexity is a property of the universe. Because the built environment 'universe' is rich and has many diverse forms and practices, it cannot be understood by using simple and usually mechanistic or linear means (Battram, 1998; Brock *et al*, 1991). Battram argues that complexity theory studies ninety five percent of situations and traditional sciences study the other five percent. The study of complex systems runs somewhat contrary to the normal (or reductionist) approach followed in physics, chemistry, biology and economics. The basic fundamentals of these disciplines are that if components of the systems are understood, one can easily formulate problems and infer consequences based on the knowledge gained from studying single components. Complex systems like the built environment cannot be understood by studying parts in isolation. The very essence of built environment complex systems lies in the interaction between parts and the overall behaviour that emerges from the interactions. The built environment (BE) systems must be analyzed as a whole. Achieving this goal will require a shift in the current built environment research. The new focus should be on coordinated built environment research through developing interoperable complex systems to address the needs for integrated systems in order to optimize and deliver future sustainable built environment.

ARCHITECTURAL ENGINEERING AND DESIGN MANAGEMENT ■ 2008 ■ VOLUME 4 ■ PAGES 3–4 earthscan
doi:10.3763/aedm.2008.S007 ISSN: 1745-2007 (print), 1752-7589 (online) www.earthscanjournals.com

IN SEARCH OF EMBRACING COMPLEXITY IN BUILT ENVIRONMENT

The papers in this issue of *Architectural Engineering and Design Management* seek to examine and to harness this fundamental paradigm shift in understanding real world complex problems. Each of the papers offers a different perspective on the nature of complexity application in the built environment sector. Allen suggests in his paper that a better understanding of complex systems is useful since the possibility of structural change, learning and innovation are considered, and that complex models provide a basis for understanding policy and, more particularly, issues such as sustainability. This is interesting since we are probably all aware of the difficulties of getting to grips with sustainability, which has been described in some quarters as being nothing more than a nebulous concept.

Bachman highlights the relationship between architecture and complexity and discusses how to bring together the complexity theory and our best understanding of how the universe really works vis-à-vis how buildings and cities are designed to become artefacts of that reality. Fernandez-Solis addresses the need to 'know why' we build by looking first at the systemic nature and complexity that informs the construction industry's current paradigm. He explores the perceptual processes of obfuscating manufacturing–industrial–engineering mindset. Rooke *et al.* explore how complexity theory may help to resolve the problems of organization through both modelling and metaphorical applications in the through-life management of the built environment. It is noted what models and metaphors have in common – both are forms of analogy. Mak *et al.* consider the building design process as a complex process that requires an integration of creativity, rationality and self organization. The authors provide an alternative framework for building design and creating a hierarchical structure of feng shui knowledge for building design.

Taken as a whole, these papers offer insights into the manner in which complexity theory can be important in shaping a new research agenda in the built environment sector. I am of the opinion that complexity has thus far been poorly exploited in built environment research. There are many implications of complexity science for BE research, but complexity theory cannot and must not be seen as the only definite explanation for elusive BE problems. We should seize the opportunity and embrace the study of complex systems. This will significantly augment the intellectual basis of the discipline and expand considerably the boundaries of the BE research agenda.

REFERENCES

Battram, A., 1998, *Navigating Complexity: The Essential Guide to Complexity Theory in Business and Management*, Spiro Press.

Brock, W. A., Hsieh, D.A. and LeBaron, B., 1991, *Nonlinear dynamics, chaos and instability: statistical theory and economic evidence*, Cambridge, MA, The MIT Press.

Dunn, T., 2003, 'Complex adaptive systems meet insurance risk', http://www.actuaries.ca/meetings/investment/2003/GS15_Dunn.pdf

ARTICLE

The importance of complexity for the research agenda in the built environment

Peter M. Allen

Abstract

Traditionally, science has attempted to understand urban systems using a reductionist approach in which the behaviour of a system (city or region) is represented as being an equilibrium mechanical interaction of its components. These components are either 'representative agents' for the different categories of supply and demand that inhabit the system, or the buildings and transport links that they create. This equilibrium assumption views their spatial distribution as optimal and stationary. Over recent decades, attempts have been made to introduce more dynamic approaches, in which equilibrium is not assumed, and there are many models and methods that try to do this. However, this still denies the essential complexity of the urban or regional system in which activities, natural endowments, culture, skills, education, health, transport, house prices and the global economy all combine to affect the evolution of the system. Just as in ecology, the key to the survival of cities and regions is the diversity, and innovative and adaptive power of people and society to counter new difficulties and create new opportunities. This fluid, adaptive power is a product of the complex system and can only be modelled and anticipated to a limited degree. However, cities and regions can limit the possibility of successful adaptation if they are too well-organized or too unimaginative. New models of adaptive organization allow us to understand better the need for integrated views linking land-use changes to environmental, socio-economic and cultural factors. These provide a new, more open way of considering the importance of adaptable, emergent networks, and the need for multiple and burgeoning accessibility to others.

■ *Keywords* – Complexity; evolution; co-evolution; urban dynamics; multi-agent models; integrated models; sustainability

INTRODUCTION

In order to make effective designs, investments and policy decisions in cities, we need to understand the multiple decisions and actions made by the multiplicity of agents and entities involved. This really means that we need to understand the options that they perceive and the trade-offs that their value systems cause them to make, and through this to know how they will react to some policy, action or investment that is contemplated. Only then would we have a reasonable basis on which to identify emerging problems and to evaluate different possible policy or decision responses.

In fact, the behaviour of *complex systems* offers an appropriate set of concepts with which to begin a new reflection on human systems. In this new view, *non-equilibrium* phenomena are much more important and offer a new understanding of the natural emergence of structure and organization in systems with many interacting individual elements. In this paper, the author briefly refers to new models of evolutionary and regional systems that show how the dialogue between the individual and collective levels generate successive spatial structures, with characteristic patterns and flows. These represent a co-evolutionary behaviour and organization beyond the 'mechanical' where the locations and behaviours of the actors are mutually interdependent, where the system has many possible responses to

doi:10.3763/aedm.2008.S907 ISSN: 1745-2007 (print), 1752-7589 (online) www.earthscanjournals.com

perturbations, and where the urban system can change, adapt and maintain rich, diverse and varied strategies. This view of suboptimal behaviours, imperfect information, mistaken inferences and the power of creativity is contrasted with the traditional mechanical representations of human systems. The models discussed here offer a new, quantitative basis for policy exploration and analysis, presenting the opportunity to take into account the longer-term implications for the system as a whole.

EVOLUTIONARY COMPLEX SYSTEMS

In a series of articles, the fundamental properties of complex systems have been established (Allen, 1983, 1990, 1997a). Let us consider the most basic problem of modelling a natural ecosystem. We can establish the different species that exist there and then find out how many of each population there are. We can also, by sampling, find out which population eats which other population and calibrate the multiple plant/herbivore and predator/prey interactions. Once this is established, we can put the whole system of equations on a computer and run it forward. What happens is shown in Figure 1.

The model collapses down to a simple, reduced structure. This is an astonishing result. It means that although the model was calibrated on what was happening at time t = 0, it diverged from reality as time moved forward. The real ecosystem stayed complex, and indeed continued to adapt and change with its real environment. But this shows us that the mechanical representation of reality differs critically from that reality. What is missing? This can be discovered if we examine carefully the assumptions that we made in formulating the population dynamics. What happened is that the loop interactions of a real ecosystem form parallel food chains, with cross connections and complications of course, but essentially with each level feeding on the lower one, some of these dying and others being eaten by the level above. The whole system of food chains loops back through death and micro-organisms that recycle all the carbon and minerals. When we run the population dynamics with the fixed birth, death, capture and escape rates that we have found on average in the real system (in analogy with chemical reaction rates), then the food chain with the highest performance simply eliminates all the others. In other words, selection between metabolic chains operates and this selects for the highest performing chain. However, reality does not. Therefore we need to understand what is missing between the dynamic model and the original real system.

The key answer is that what is missing is the *internal diversity of the populations*. In chemistry, one molecule is very like another, and the only difference is their spatial location. Dissipative structures can create spatio-temporal patterns because of this. But

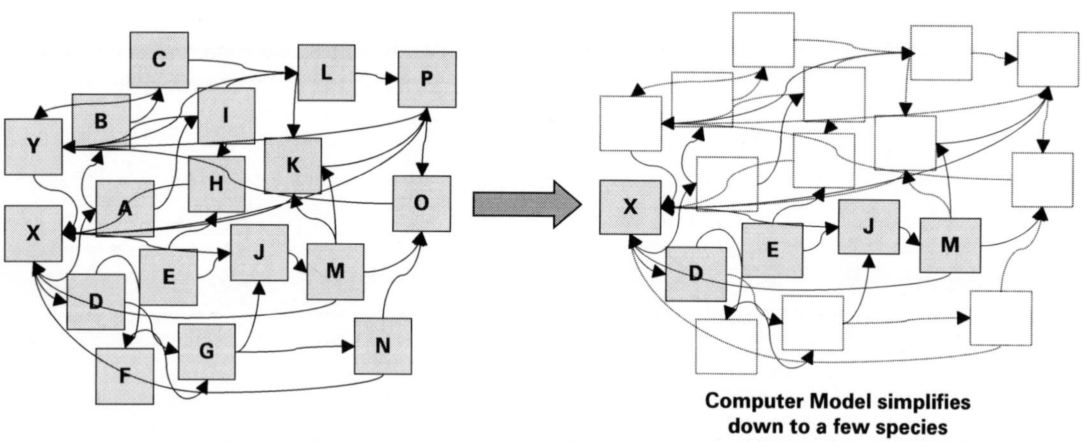

FIGURE 1 A calibrated ecosystem represented by the population dynamics of its constituent species collapses when run forward in time

populations of organisms differ in an infinite number of ways. First, in location, but also in age, size, strength, speed, colour, etc. and so this means that whenever a population, X, is being decreased by the action of some particular predator or environmental change, then the individuals that are most vulnerable will be the ones that 'go' first. Because of this, the parameter representing the average death rate will actually change its value as the distribution within the population X increases the average 'resistance'. In other words, the whole system of populations has built in, through the internal diversities of its populations, a multiple set of self-regulatory processes that will automatically strengthen the weak and weaken the strong. In the same way that reaction diffusion systems in chemistry can create patterns in space and time, so in this more complex system, the dynamics will create patterns in the different dimensions of diversity that the populations inhabit. But neither we, nor the populations concerned, need to know what these dimensions are. It just happens as a result of evolutionary dynamics.

In this case it becomes key to understand the sequence of assumptions that take us from reality to a mechanical representation of that reality (see Figure 2). Table 1 sets out the kind of models that result from a particular set of assumptions.

This succession of models arises from making successive, simplifying assumptions, and therefore models on the right are increasingly easy to understand and picture, but increasingly far from reality. They also are shorn of their capacity to evolve – their real underlying exploratory, error-making processes. The operation of a mechanical system may be easy to understand but that simplicity has assumed away the more complex sources of its ability to adapt and change. They become more like 'descriptions' of the system at a particular moment, but do not contain the magic ingredient of micro-diversity that will really allow the system to undergo structural change and create a new, qualitatively different system, with some new variables and some emergent performance. The ability to adapt and change is still present in the 'evolutionary' model that only makes assumptions 1 and 2, but not

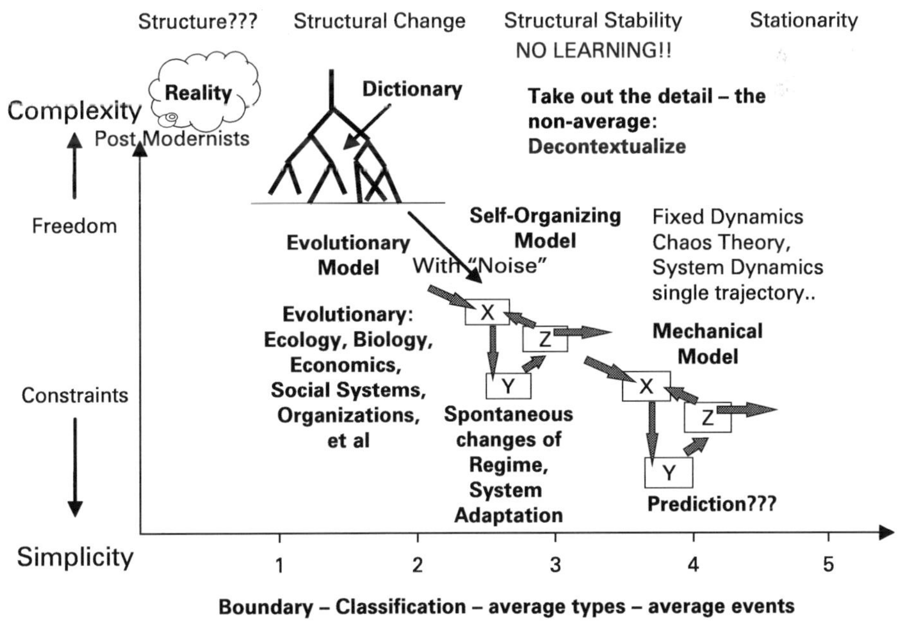

FIGURE 2 This shows the results of successive simplifying assumptions that take us from a complex evolving system to its mechanical representation

TABLE 1 The different kinds of mathematical model that result from successive assumptions

NUMBER	ASSUMPTION MADE	RESULTING MODEL
1	Boundary assumed	Some local sense-making possible – no structure supposed
2	Classification assumed	Open-ended evolutionary models – structural change occurs
3	Average types	Probabilistic, non-linear equations – assumed structurally stable
4	Average events	Deterministic, mechanical equations – assumed structurally stable

those of average type and average behaviours. This, therefore, tells us that the evolutionary capacity is generated by the behaviours that are averaged by assumptions 3 and 4 – average types and average events – and therefore that organizations or individuals that can adapt and transform themselves do so as a result of the generation of micro-diversity and the interactions with micro-contextualities. This tells us the difference between a reality that is 'becoming' and our simplified understanding of this that is merely 'being' (Prigogine, 1980).

In reality, complex systems thinking offers us a new, integrative paradigm, in which we retain the fact of multiple subjectivities, and of differing perceptions and views, and indeed see this as part of the complexity and a source of creative interaction and of innovation and change. The underlying paradox is that knowledge of any particular discipline will necessarily imply 'a lack of knowledge' of other aspects. But all the different disciplines and domains of 'knowledge' will interact through reality and so actions based on any particular domain of knowledge, although seemingly rational and consistent, will necessarily be inadequate.

Management or policy exploration requires an integrated view. These new ideas encompass evolutionary processes in general and apply to the social, cultural, economic, technological, psychological and philosophical aspects of our realities. Often, we restrict our studies to only the 'economic' aspects of a situation, with accompanying numbers, but we should not forget that we may be looking at very 'lagged' indicators of other phenomena involving people, emotions, relationships and intuitions, to mention but a few. We may need to be careful in thinking that our views will be useful if they are based on observations and theories that refer only to a small sub-space of reality – the economic zone.

The underlying causes and explanations may involve other factors entirely and the economic 'effects' of these may be delayed, ripples or possibly tidal waves. What matters over time is the expansion of any system into new dimensions and conceptual spaces as a result of successive instabilities involving dimensions additional to those the current 'system' appears to occupy.

Rational improvement of internal structure – the traditional domain of systems thinking – supposes that the system has a purpose and known measures of 'performance' that can indicate the direction of improvements. But this more fundamental structural evolution of complex systems that results from successive invasions of the system by new elements and entities is characterized by emergent properties and effects that lead to new attributes, purposes and performance measures. In the next sections, therefore, we attempt to show that this structural evolution is not in fact 'random' in its outcome, since successful invasions of a system are always characterized by the revelation of positive feedback and synergy, creating particular, new, internally coherent structures from a growing, explosively rich set of diverse possibilities.

The realm of complex systems models that we wish to develop aim to make only the first two assumptions and to study cities and regions as evolving, self-transforming systems in which behaviour, decisions and the value systems underlying these all evolve over time. This leads to a view of a city or region as a complex evolution of spatially distributed learning reflecting local stresses, opportunities and exploratory responses such that people not only change what they do, but also their knowledge of what they could do and what they want to do. Qualitative, structural changes occur both in the macroscopic forms of the collective structure and also in the microscopic structures within individuals' brains that govern their trade-offs and decision making, which in turn govern the future structural evolution of the collective system and of the individuals that inhabit it.

EVOLUTION OF THE BUILT ENVIRONMENT

The built environment is really a multi-layered historical patchwork of structures and links that express individual and contextual requirements at a given time, but which shape the evolution of the future in a typical 'path-dependent' process of complex evolution. In reality a city is a complex system, as is a neighbourhood, a block, a household and an individual. These represent nested levels of description and we can develop mathematical models that can explore different possible evolutionary pathways and possible futures under the assumptions of different possible interventions. This work started in 1976 when the US Department of Transportation commissioned our early research on developing such models, initially only dynamic, but later on developing fully complex, 'learning' multi-agent models. The essence of these models is shown in Figure 3 in which the locational patterns of people, jobs, transport and infrastructure are coupled together, so that their combined evolution can be explored under different interventions and plans. The key idea is that changing transport costs or access changes the locational choices of people and activities, which in turn changes the patterns of traffic flow and the locational choices of people and activities, and so on. These strong feedback relations mean that the system is unstable and can exhibit different possible trajectories into the future. In examining possible decisions, therefore, we need to see whether the different possible trajectories of the system are all acceptable or whether there is a probability that some very unpleasant outcomes may be possible.

An important point to underline concerns the reasons for which we travel at all.

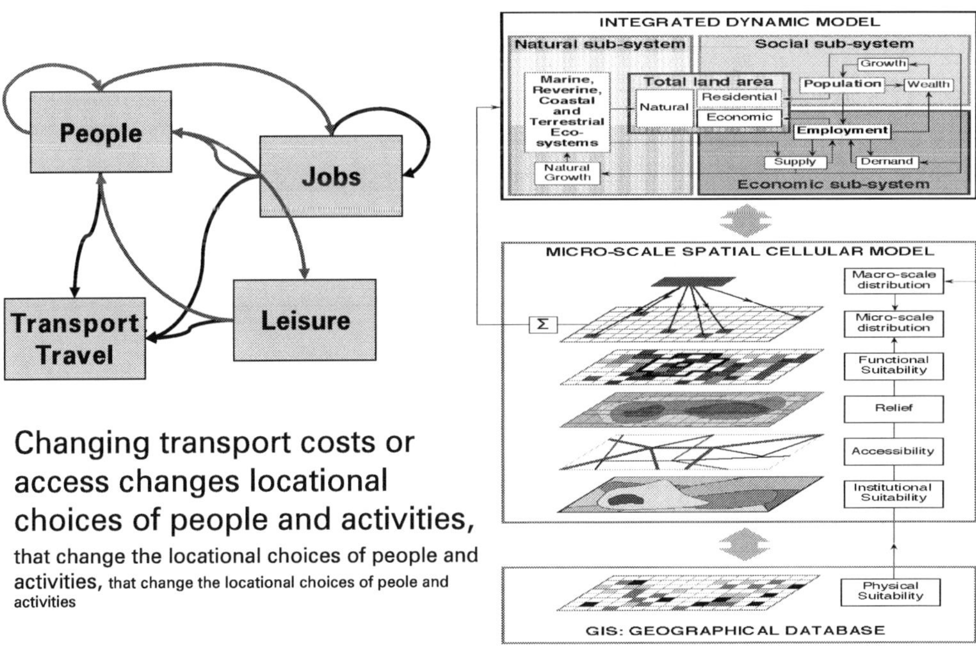

FIGURE 3 Software systems have been developed that allow the interacting spatial distributions of people, jobs, leisure facilities and transportation to be studied (White and Engelen, 2001)

If we ask why people travel at all, we find that it is because of the spatial distribution of diverse activities and opportunities:

- dispersed distribution of affordable/desirable housing
- concentrated distributions of employment
- concentrated distributions of retail opportunities
- dispersed distributions of leisure and cultural facilities.

Transport demand is therefore generated by these spatial distributions and an important point is that these distributions are all *co-evolving* with each other over time and also reflect changes in the transportation systems. In short then, the demand for transport is generated by the details of the different distributions that affect each other, and the transport congestions and patterns of access then affect the locations of the different spatial distributions, which in turn feed back on the demand for transport. This is particularly evident at the present time in the UK when spatially dependent house price rises are shaping longer commuting patterns for large numbers of people and even threatening the successful functioning of cities as ordinary workers, particularly in the public sector, find it increasingly difficult to find homes within reasonable distances of their work.

In view of this complexity, and the intertwined effects of transportation and spatial structure, it seems clear that there is a problem for the evaluation of transportation policies and plans. How can there be an overall assessment or evaluation of any plan for new roads or for public transport systems unless these complex effects are assessed? The answer is that, in fact, they are not. Decisions concerning urban highways, new tram and metro systems are really decided on the basis of politics and fashion.

Since the 1970s, work has been going on that attempted to develop computer models that would take into account the complex interactions of linked responses that lead to a co-evolution of urban structure (patterns of retail, commercial and manufacturing employment, and different qualities of residence) with transportation infrastructure. These models are based on the following characteristics:

- different types of actor at each zone, with characteristic needs to be fulfilled
- these characteristic needs are stable, but the behaviour of actors depends on the changing circumstances
- the spatial distributions of the different types of job and different kinds of people affect each other as the potential for housing demands, commercial activities and for travel affect and are affected by transportation and land use.

The development of these models has been described by Allen (1997b). After an initial phase that developed models suitable for some US and European cities, an example based on Brussels was developed to demonstrate the potential utility of the approach. The model represents the interacting behaviours of the actors in the urban system in several ways – as they each modify their behaviour as a function of the changing opportunities and pressures, as they each pursue their own goals for the location and relocation of employment according to the functional requirements and, as private citizens, as a function of their means and opportunities. The spatial dynamics can therefore generate and capture the complex effects of housing price dynamics and also the complex effects of planning regulations on commercial and industrial employment, as well as the effects of changes in the transportation systems.

In Figure 4 we see the interaction diagram of the different types of actor considered adequate to represent the spatial evolution of a city like Brussels in the 1980s. It has different possible interaction mechanisms between them which express the need for flows of goods, services and people between different locations and also the pressure of spatial concentration affecting land prices and rents.

The mechanisms above, when run under a scenario of overall growth, spontaneously generate self-consistent urban spatial structure for the seven types of actor, as well as the corresponding flows of goods, services and people.

Such a model can therefore be used to explore the effect on the spatial structure of possible modifications of the transportation system. This could correspond to plans for new roads, tramways

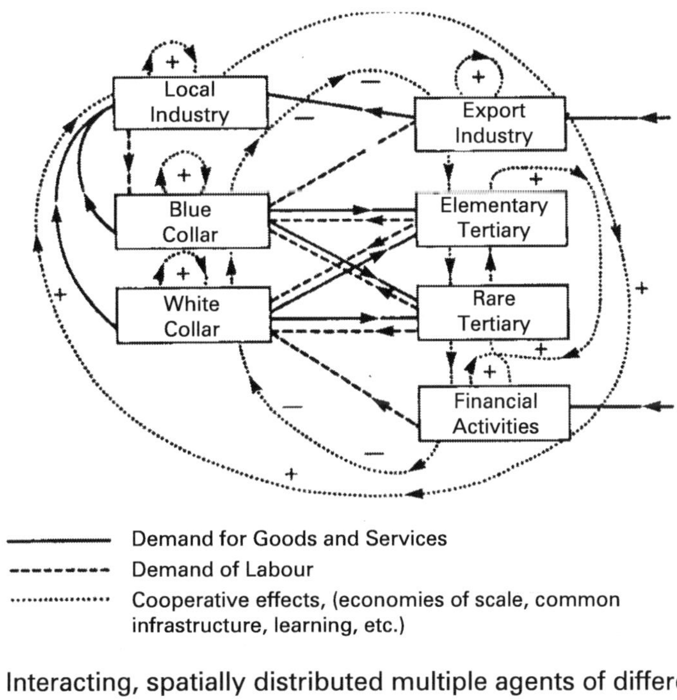

FIGURE 4 The interaction diagram that can provide the essential mechanisms operating within the city

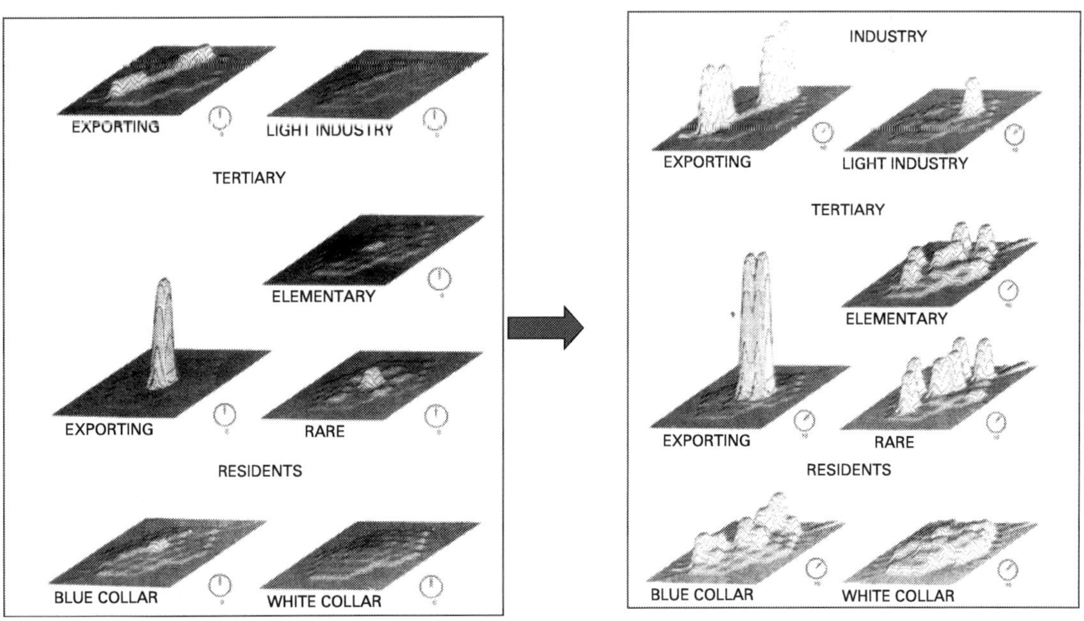

FIGURE 5 Emergence of complementary spatial structure of the seven variables

or metro system. By looking at the changes in structure that follow some intervention, our model can explore the impacts over time of a given action, as actors respond to the new situation, and their changed behaviour affects other actors in turn, creating a complex spatial multiplier (see Figure 6).

This allows us to examine the complex effects of the cascading interactions under different possible plans for the metro – possible routes, locations of stations, train sizes and frequencies. While not pretending that each outcome is a real prediction that is accurate, what matters are the relative differences between simulation outcomes, since these will show the 'relative effects' of different routes, of more or less trains, etc. Similarly, the model can be used over the longer term to examine some strategic issues such as the effect on decentralization–centralization. This surely is one of the major questions that affects any city – will this action influence the existing trends and patterns of migration of jobs and people into the periphery? Clearly, the action of building metro systems is one that tends to 'allow' people to travel to the central part of the city with some ease. By the additional use of 'park and ride' car parks at edge-of-town metro systems, it may even encourage the further out-migration of residents from the city, but anchor employment at its centre. The point about this is that any evaluation of plans for a metro system needs to include not only the projected costs of the system, but also the projected effects on the city. These projected effects cannot be calculated simply from the expected 'traffic' that switches from roads to the metro, but also need to encompass the spatial changes brought about to the residential, commercial and employment sectors. In particular, it is important to have some idea of the strategic impacts of a transportation scheme and whether it will tend to accelerate or reverse some basic trends that are running in the system at present.

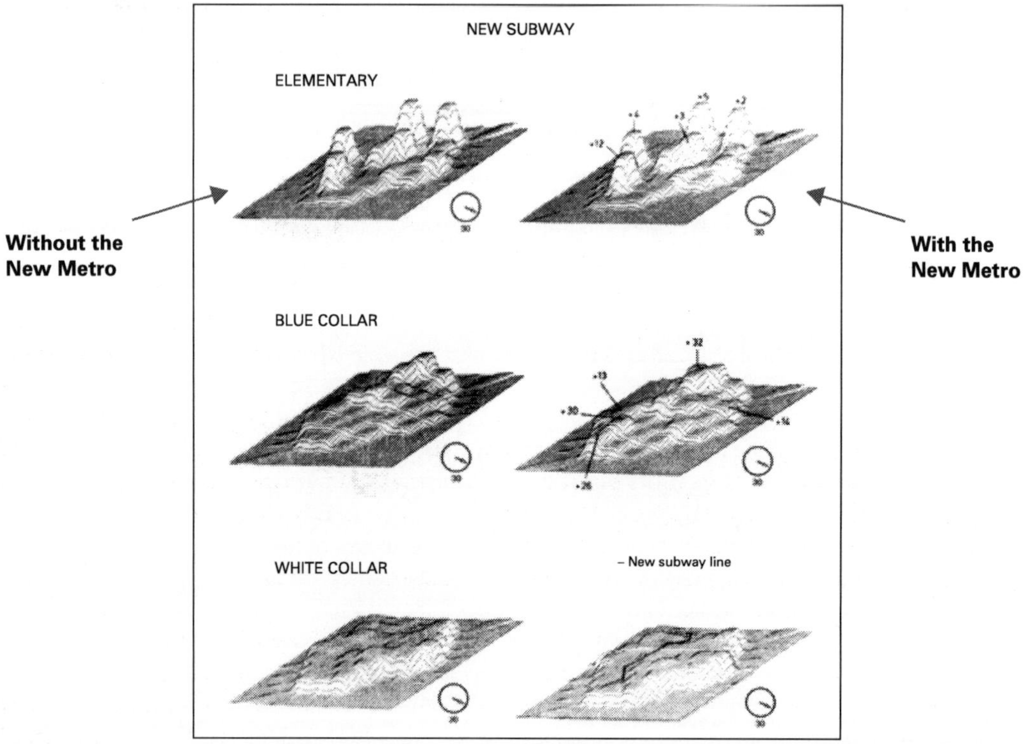

FIGURE 6 The distributions of residents and of tertiary activity are shown for simulations with and without a new metro system

Despite being developed over a decade ago, models of the kind described above are still not used by decision makers such as regional, urban or local authorities. It seems clear then that such decisions are left to the intuitive judgement of such authorities, acting under the advice and pressure of competing lobbies and groups with particular interests. This cannot be a good way to make decisions. In particular, it seems evident that some method is required for estimating the strategic impacts on the growth patterns of the city, effects on house prices, on residential and commercial development and, in turn, on future traffic patterns, energy consumption, pollution, etc. The framework discussed briefly here is a candidate for this and therefore deserves a renewal of interest in its development and adoption.

CONCLUSION

The conclusion of this contribution is that the new ideas emerging from complex systems thinking offer us a new basis for understanding and living in the real world. Since the possibility of structural change, learning and innovation are considered, these kinds of models provide a new basis for policy exploration, particularly with respect to issues of 'sustainable' development. In these, the 'bio-physical' part of the system (the hydrology, soils, vegetation, ecology, physical infrastructure, etc.) is linked dynamically to the 'human' part of the system that is driving the exploitation of resources, both natural and human.

These developments underline the fact that these models should not be thought of as only of 'academic' interest. Nor are they just biological or chemical metaphors. The fundamental points that have been made concern the scientific basis of understanding. Understanding is achieved in a trade-off between simplicity and realism. The whole question is whether or not a simple enough description can be found that is still sufficiently realistic to be useful. In the past, the desire for tractability has led to the use of very strong assumptions such as that of 'equilibrium', which is necessary prerequisite for a normal 'cost/benefit' analysis of a decision. It is the author's contention here, that such methods are incorrect – although possibly better than nothing. The new methods presented here are still not used operationally, which means that any strategic aims that are involved in the decision to invest in new transportation systems are really based on the personal intuition of the people involved. Of course, these can be correct, but in general it would be good to be able to provide better information about the probable consequences of such schemes.

The complex systems view of the built environment is that it should be an emergent artefact that responds to people's needs in a changing world. It underlines the importance of emergent collective processes, requiring greater social responsibility between creators and builders and end users. Instead of taking each building project individually and responding to the minutiae of market forces, we need to conceive of towns and cities in a people-centric holistic way, in order to meet the major challenges facing society.

Worldwide demographic growth is creating the urgent need to implement the concepts of sustainability in our built environment and to deal with a changed demographic profile in many countries, where older people will make up an increasing part of the population. Construction has a pivotal role to play in preserving the environment, mitigating existing polluted areas, saving natural resources such as water, energy and raw materials. In particular we shall need to change radically the numbers of people that work far from where they live and this will require a major shift in the design of towns and cities. We must move away from the 'dormitory-commuting-and-business' model that has become the norm, and we need to rediscover some modern 'medieval' type of urban system that will again allow communities and local activities to re-emerge. Clearly ICT can greatly reduce the need for daily commuting to distant workplaces and so new lifestyles can become possible. It is important that the construction industry, architects and developers respond to these new needs and opportunities.

Obviously, climate change is creating the vital need for change in the built environment as we try to adapt our homes and workplaces to a more variable climate. We must also consider the increased need for better infrastructure to control increasingly frequent natural hazards (such as floods, landslides and earthquakes).

Quality of life is something that the built environment should be providing and to this end we need precisely the integrated view that comes from complex systems thinking. Instead of separating home, work, transport, health, education, communications, natural environment, food and water, the complex systems approach treats them as a coupled whole and allows us to reflect on the overall, multi-dimensional impacts on our lives of particular investments or innovations. This is desperately needed and difficult to implement with our fragmented governance and competing ministries. Furthermore, reliance on the 'market' will not achieve it either since this will always tend to adopt short-term profit as its leitmotif, reflecting the values of the anonymous managers of money and their investors who are judged purely on their financial success. The integrated, sustainable, complex view needs to be carried forward by those that care about where and how they live and work, and the quality of life offered by society.

The history of a successful society within a region is largely a tale of increasing cooperation and complementarity, not competition. An economy is a 'complex' of different activities that to some extent 'fit together' and need each other. Competition for customers, space or for natural resources is only one aspect of reality. Others are of familiar suppliers and markets, local skill development and specialization, co-evolution of activities to each other, networks of information flows and solidarities, that lead to a collective generation and shaping of exchanges and discourse within the system. Evolution is not about a single type of behaviour 'winning' through its superior performance, but rather by increasing diversity and complexity. The models we propose are therefore ones that can help us deal with the overall, integrated effects of the coupled decisions of multiple actors, allowing us better insight into the consequences of possible policies and actions.

AUTHOR CONTACT DETAILS

Professor Peter M. Allen: Complex Systems Research Centre, School of Management, Cranfield University, Bedfordshire MK43 0AL, UK. Tel: +44 (0)1234 754800, fax: +44 (0)1234 754804, e-mail: p.m.allen@cranfield.ac.uk

REFERENCES

Allen P.M., 1983, 'Self-organization and evolution in urban systems', in R. Crosby (ed), *Cities and Regions as Non-Linear Decision Systems,* AAAS Selected Symposium 77, Boulder, US, Westview Press.

Allen, P.M., 1990, 'Why the future is not what it was', in *Futures*, 22(6), 555–569.

Allen, P.M., 1997a, 'Cities and regions as evolutionary complex systems', in *Geographical Systems*, 4, 103–130.

Allen, P.M., 1997b, *Cities and Regions as Self-Organising Systems: Models of Complexity,* London, Taylor and Francis.

Allen, P., Sanglier, M., Engelen, G. and Boon, F., 1985, 'Towards a new synthesis in the modelling of evolving complex systems', in *Environment and Planning B: Planning and Design*, Special Issue, 12(1), 65–84.

Prigogine, I., 1980, *From Being to Becoming: Time and Complexity in the Physical Sciences*, San Francisco, Freeman.

White, R. and Engelen, G., 2001, 'High-resolution integrated modelling of the spatial dynamics of urban and regional systems', in *Computers, Environment and Urban Systems*, 24(5), 383–400.

ARTICLE

Architecture and the four encounters with complexity

Leonard R. Bachman

Abstract

In the surge of scientific revelations concerning chaos, cybernetics and systems theory since the Second World War, architecture has been drawn into several encounters with dynamic complexity. Architects have often danced with this complexity, but there has never been a full embrace – the encounters only amounting to skirmishes between two radically different perspectives. Increasingly however, these two histories of design and science are convergent, drawn together by forces of post-industrial knowledge, society, globalization, economic interdependence and ecological sustenance. Through past literature, that convergence can be framed by four modes of architectural complexity – wicked, messy, ordered and natural – but the history of these encounters has seldom been couched in terms of their teleological and dynamic influences, nor have their separate operations been characterized or connected in a way that adds coherence. This paper addresses this missing history by making explicit distinctions between each of the four modes of architectural complexity, tracing their early proponents and characterizing their operations.

■ *Keywords* – Complexity; design; history; systems; wicked; messy; ordered; natural; post-industrial; Gestalt

INTRODUCTION

Complexity is the way in which a whole is different from the composition of its parts. (M.H. van Emden, 1971: 7)

Van Emden's description of complexity provides a particularly appropriate filter for discussing complexity in architecture. While similar to other such definitions, the link to Gestalt theory and Aristotle's 'the whole is more than the sum of its parts' parallels our understanding of how important buildings emerge from piles of bricks and glass. Van Emden's passage is used here to qualify and catalogue the manner in which design complexities have been encountered by architects. Once identified, the different dimensions of these encounters can be slotted into the missing history of how architecture has begun to frame wicked problems, deep interactions and organizational dynamics of building design. Indeed, this is a missing history precisely because those wicked, interactive and dynamic factors have often been encountered in architectural thinking but without any coherent connection between their underlying indeterminacy and the methods that architects use to understand them. In short, this paper is a history that connects architectural views of complexity to scientific ones.

In the surge of revelations concerning complexity, cybernetics and systems theory since the Second World War, architects have clearly come to contemplate design complexity with a different attitude to that of scientists. The literature of architectural thought and the built works it produces has continually encountered complexity theory, but there has never been a full embrace of what science tells us is our best understanding of how the universe really works vis-à-vis how buildings and cities are designed to become artefacts of that reality. Increasingly, however, these two histories are convergent, drawn together by networking forces of post-industrial knowledge, society, learning institutions, globalization, economic interdependence

doi:10.3763/aedm.2008.S407 © 2008 Earthscan ISSN: 1745-2007 (print), 1752-7589 (online) www.earthscanjournals.com

and ecological sustenance. Perhaps some recognition of the differing mindsets as described in this paper will advance their coming together.

RATIONALE

Architecture is a continuous encounter with complexity, an encounter that is inherently manifested by concerns such as initial project requirements, design goals, context of place and performance in use. These complexities can be sketched out as:

- Design begins with a 'bottom-up' need to collect and deductively digest vast amounts of information related to user needs, site conditions, code requirements and so forth (Peña, 1969).
- This bottom-up need pairs directly with an inductive 'top-down' ambition for the design outcome as an authentic member of a place, a building that both has its own identity and its role in the cultural fabric (Jacobs, 1961).
- Next, architectural design must accommodate different sets of stakeholders and facilitate collaboration as well as be robust in the face of inevitable changes in its use, maintenance and control (Simon, 1969).
- Finally, the occupied building must be conceived as a complex organism with dynamic interactions, operations and well-balanced flows of everything from people to daylight, information to heat, and merchandise to air (Ackoff, 1968; Groák, 1992).

These are not just a sequence of four complicated and detail-heavy tasks that the architect must reduce and tame into manageable problems so that they are suitable for intuitive and imaginative design responses; rather, they are separate dimensions of what can be seen as the possibilities of fully mindful design. This mindfulness, in turn, is what produces animated buildings that respond to their own unique essence, fit their environment and are adaptive to change. From the perspective of complexity science, this mindfulness represents the cybernetic, teleological and dynamic systems behaviour of buildings as organisms. To architects, this mindfulness represents buildings that are authentic, elegant and timeless.

Various aspects of architectural complexity have been previously addressed by several thinkers who will be important to how this discussion is developed (Figure 1). Our present-day understanding of architectural design in the context of post-industrial society certainly engages their writings from several viewpoints – sustainability, strategic planning, post-occupancy evaluation, continuous commissioning, as well as a host of other sophisticated investigations and accountabilities coming to the fore.

Despite those previous works, however, critical distinctions of architectural complexity have not been well enumerated; and in much architectural practice they are too frequently lumped collectively into work that can presumably be reduced to intuitive solutions. So instead of recognizing complexities, design investigation is generally treated as the normative problems of programming, site design, interiors and service systems. Unfortunately, these fallback normative approaches are reductive and procedural-based activities and are organized around straightforward methods that often miss the systemic complexity underlying a design challenge. This approach has been characterized as a futile attempt to 'tame complexity' (Rittel and Webber, 1973). As Churchman (1967: B141) points out in an introduction to some of Rittel's early work, this tact is inherently deceptive and morally wrong. He would rather that we say: 'Look, I've tamed not the whole problem, just the growl; the beast is still as wicked as ever.'

Beyond normative design practice then, in the more radically artistic or literary models of architectural design theory, the complex underpinnings of the design project may be seen as so daunting that they are simply denied or ignored in favour of purely subjective or intuitive approaches. Hence, the most complex, vital and unique essence and therefore authentic opportunity of the design challenge can be disguised, corrupted or completely lost. To quote Christopher Alexander's *Notes on the Synthesis of Form* (1964: 1): '...when a designer does not understand a problem clearly enough to find the order it really calls for, he falls back on some arbitrarily chosen formal order. The problem, because of its complexity, remains unsolved.'

To summarize this critique, we can say that architectural design has generally treated the deeply interrelated and dynamic characteristics of buildings as if they could be reduced to mechanistic problems.

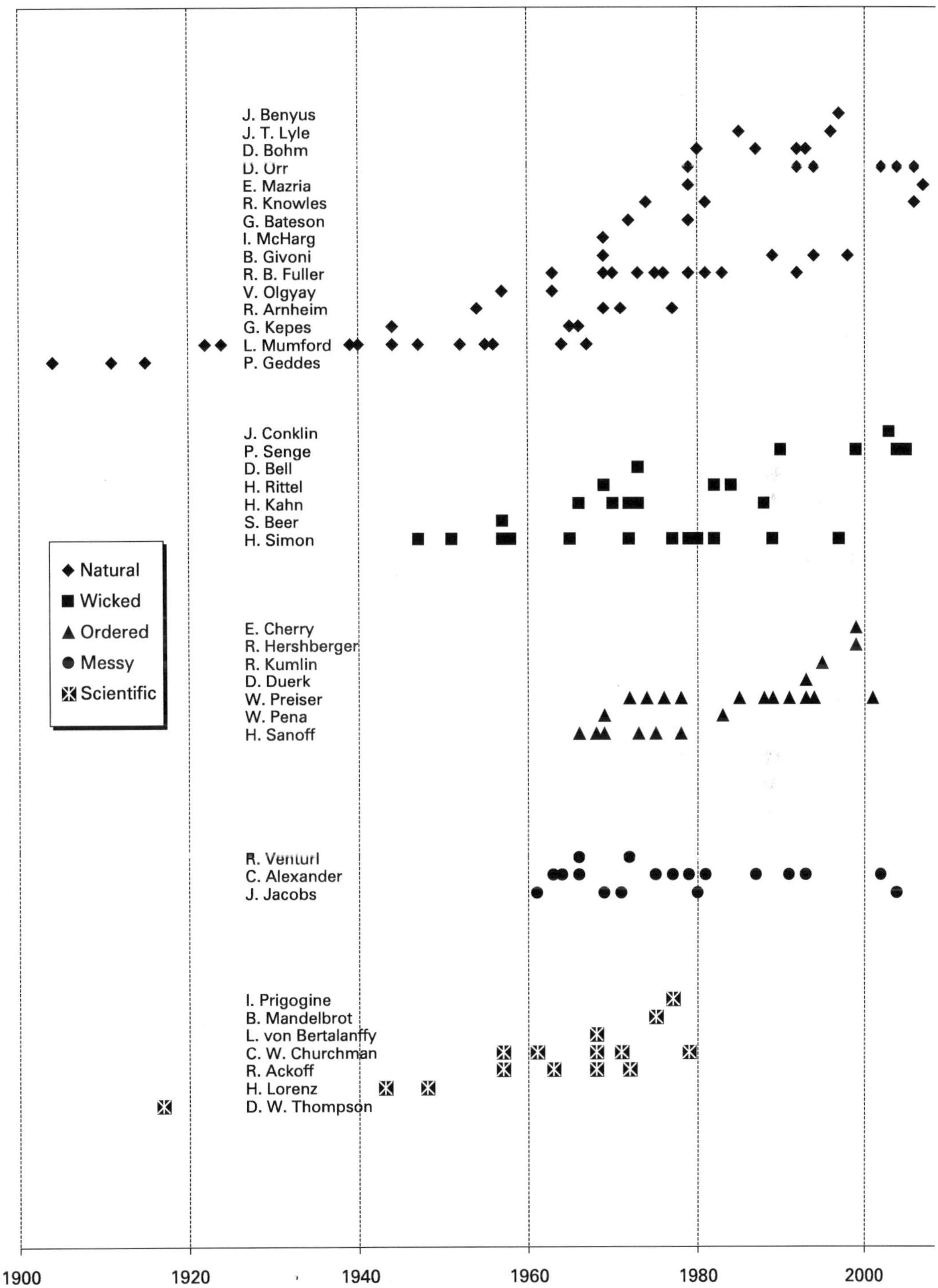

FIGURE 1 Timeline of complexity in architectural literature with some corresponding works on the science of complexity

In turn, these mechanistic problems are presumably to be solved by a normative distilling of data down to whatever intuitively seems to matter most, or else by ignoring the complexities altogether in favour of purely formalistic solutions. At best, the detail level of complexity may be recognized as an issue of complicated information, but the dynamic aspects of interaction, feedback, collaboration and integration are still too often sublimated. It is not that architectural thinking has never encountered complexity, but rather that the normative level of practice avoids, denies or dodges the issue; and thus the critical level of design thinking has yet to fully acknowledge the potential of what is latent in its own literature.

This paper sets out an argument for the deeper principles of four different modes of complexity as they have actually been encountered in architecture. While those various arguments have been opened before, the history of their emergence has seldom been couched in terms of their indeterminate, teleological and dynamic influences; nor have their separate histories been connected in a way that adds coherence. The present work addresses this missing history by making explicit distinctions between each of four complexity modes, tracing their early proponents and characterizing their separate operations. The primary intent is to establish a meaningful taxonomy of the different encounters architecture typically has with complexity and to show accordingly that each has a separate but heretofore largely undesignated history within the discipline. The ultimate goal then is to enhance the richness of these encounters by correctly placing them inside the imaginative aspect of architectural design, rather than leave them as undesignated routines of mechanistic data reduction.

THE FOUR ENCOUNTERS WITH COMPLEXITY

The scientific study of complexity generally refers to dynamic complexity where deep interrelations result in cause–effect events that are not directly observable or intuitively straightforward. For the purposes of this discussion, complexity includes the study of non-linear systems, cybernetics and chaos. This differentiates dynamic complexity from the tamer pursuits of detail complexity, which is merely about complicated information or problems of combination or optimized logistics. In short, detail complexity is about mechanistic problems with simple solutions while dynamic complexity is about systems and systemic solutions. We can distinguish between them by comparing a tree farm to a forest, or complex living organisms to Frankenstein's piecemeal artifice of a monster.

In this scientific sense of the term then, complexity has seldom been directly acknowledged in built or written works of architecture. The exceptions are rare and incomplete in their treatment (e.g. Jencks, 1995). There is, however, a literature of architectural thought that borders on the topic and thrusts towards its centre, and this literature does go somewhat further than is apparent from most architectural discussion. What remains is to recognize how these architectural themes on complexity have emerged, how they are situated in the discourse of architecture, and how they may be brought to centre stage. In short, the missing history of this literature can be seen as a whole to connect architect's designerly thinking with the complex systems view of science.

By collecting the main architectural writings on complexity and organizing them according to their themes, four different architectural encounters with complexity can be described. The 'four encounters' that follow are intended to illustrate the different dimensions in which architecture has embraced complexity in that dynamic way that recognizes an underlying systemic nature. These encounters all seem to have been collectively recognized in the early and mid-1960s as clearly emergent issues and frameworks for designerly thinking. The actual events of these encounters overlap considerably, as does the thinking of those who led the charge. Indeed, some of the classifications offered in this paper are admittedly open for further argument as to which authors should be put where. To that end, some connective linking between the different encounters will be elaborated as each theme is discussed. For discussion purposes, however, the encounters are represented below by a bibliographical sketch of first edition seminal works which each, at least arguably, belong to one of the four categories of architecture's encounters with complexity.

ENCOUNTER ONE: WICKED COMPLEXITY OF THE NETWORK

The wicked dimension of complexity is defined from one side by the limits of human cognition. The other side is that of problems whose bounds are fuzzy and indeterminate, but always made of interrelating parts. Taken together, we have the compounding condition of unstructured limitless information being processed through a narrowly constricted human processor. This complexity is emphasized by recognition of the social dilemmas in which they are set, and by the post-industrial transformation to the primary production of value through knowledge manipulation and learning (Bell, 1973).

Herbert Simon (1957, 1976) is well noted for his contributions on the practical limits of human cognition. Bridging political science, psychology, cybernetics, organization theory, economics and parallel fields, he received many honours including the Turing Award (1975), The Nobel Prize for Economics (1978) and the National Medal of Science (1986). His writings on 'bounded rationality' and 'satisficing' concerning our limited ability to collect and use data are definitional. According to Simon, we eventually run out of time, patience or other resources for any project and must settle for 'satisficing' so that our solution is good enough to satisfy sufficiently ('satisfice') without actually being able to cognitively maximize the outcome of complex tasks. This fundamental wickedness of restricted human throughput is thus established by Simon's important concepts. While Simon's interests were predominantly linked to organizational theory and economics, architecture is concerned with the same cognitive limitation, especially in the collection and use of project data, building codes, site constraints, budget targets and user needs.

On the other side of wicked complexity is the formulation of design as a problem of indeterminate and unbounded information. To this end, Rittel and Webber published their landmark essay on wicked problems in 1973 – *Dilemmas in a General Theory of Planning*. They described 'wicked problems' as a class of problems distinct from simple or 'tame' problems. Some traces of Rittel's work in this area go back to at least 1967 (Churchman, 1967) and the basic idea has been traced to Karl Popper. Generally, however, Rittel is credited with establishing wicked problems as a new second generation of design approaches. The first generation is characterized as a closed and linear approach to problem solving, whereas the second generation is one of open enquiry and argument. Without recounting all 10 of Rittel and Webber's characterizations of this wickedness, it can be said that in this class of problems there are no defined starting points, operational rules, ending points or universally accepted definitions. There is never a best solution, just a better or worse one. Designers never finish, they just run out of time or other resources. In fact, every attempt to solve the problem dynamically changes the designer's previous understanding of its definition. This is certainly a good description of the design activity in any domain, and it obviously applies well to architecture and planning. 'Wicked problems' are also a close parallel of indeterminate, non-linear problems – that is to say of 'systems'. In our post-industrial knowledge production therefore, it turns out that dealing with wickedness as a design activity is fundamental to the value of design. Design and wickedness now go hand in hand.

Aside from the term's origin in the writings of urban planning, the wicked problems construct has since garnered considerable attention and use in architecture. Wickedness is cited as a term in a broad spectrum of literature on the nature of design (e.g. Buchanan, 1995). Additionally, Ackoff's *Redesigning the Future* (1974) framed the idea of wicked problems as 'messes' and Horn (2001) calls them 'social messes', thus connecting wicked problems back to the physical and social sciences.

In combination, Simon's 'satisficing' and Rittel's open-ended problems describe wicked complexity in architectural design as an attempt to apply intelligent restraint in the encounter of indeterminate challenges that inhabit a poorly bounded problem space. One has to be smart enough to know when complexity can be resolved and patient enough to work through its matrix of information and relations. Wicked problems are fundamentally a problem related to data, conversion of data to strategic information, and the use of that information by design decision-makers with limited cognition. Wickedness is thus not an essential problem of form making in design, except for how human intelligence becomes embodied in a built work.

Finally, wicked complexity needs to be set in the larger historical and global context that frames its formulation, namely Daniel Bell's *The Coming of Post-industrial Society* (1973). Bell places the US's transition to an information society somewhere in the early 1950s. This seems to portend the late 1950s and early 1960s chronology of architectural works on complexity across all four themes, but given the connection between the information society that Bell describes and the wickedness of information and bounded rationality, it seems especially appropriate to recognize Bell in the context of this first encounter.

ENCOUNTER TWO: MESSY COMPLEXITY OF THE WHOLE

Messy complexity is taken as a term from Robert Venturi's *Complexity and Contradiction in Architecture* (1966). In this text, generally recognized as a manifesto of postmodern architecture, Venturi advocated that architecture should grapple with the messy and difficult whole:

> *I am for messy vitality over obvious unity ... I am for richness of meaning rather than clarity of meaning ... I prefer 'both-and' to 'either-or', black and white, and sometimes gray, to black or white ... But an architecture of complexity and contradiction has a special obligation toward the whole ... It must embody the difficult unity of inclusion rather than the easy unity of exclusion. More is not less. (Venturi, 1966: 22–23)*

The urban thinker Jane Jacobs (1961, 1969) weighs in here as well. In her terms, an authentic belonging and sympathy to context are essential to the complex whole of design. Her work rejects the utopian regulations and optimized land planning principles that are often associated with weaker applications of modernism.

> *There is a quality even meaner than outright ugliness or disorder, and this meaner quality is the dishonest mask of pretended order, achieved by ignoring or suppressing the real order that is struggling to exist and to be served. (Jacobs, 1961: 15)*

> *To see complex systems of functional order as order, and not as chaos, takes understanding. The leaves dropping from the trees in autumn, the interior of an airplane engine, the entrails of a dissected rabbit, the city desk of a newspaper, all appear to be chaos if they are seen without comprehension. Once they are understood as systems of order, they actually* look *different [original emphasis]. (Jacobs, 1961: 376)*

> *The results of such profound confusion between art and life are neither life nor art. They are taxidermy. (Jacobs, 1961: 373)*

Or, as Peter Laurence summarizes in his detailed treatment of the Jacobs and Venturi tie:

> *When Robert Venturi began writing Complexity and Contradiction in Architecture in 1962, Jane Jacobs's* The Death and Life of Great American Cities *(1961) was the subject of dinner-table conversation across the country. Apart from Jacobs's direct influence, among the parallels that emerged from their Townscape affiliations and other interests, the concept of complexity stands out, with Venturi's conception echoing Jacobs's historic introduction of complexity science into architectural and urban theory. While Venturi did not seek to extend the scientific research put forward by Jacobs, favoring instead the complexities suggested by Gestalt theory and New Criticism, he cited similar scientific sources and derived similar principles for urbanism, as revealed by his Copley Square project of 1966. (Laurence, 2006: 49)*

Mathematician and architect Christopher Alexander (Alexander, 1964, 1966; Alexander *et al*, 1977) also shares in this perspective of messy complexity. To him there is an authentic natural order, self-emergent from spontaneous responses to prototypical situations. His differentiation, close to that of Jacobs, is between artificially contrived forms that are to a large degree invented versus naturally emerging forms that are more or less discovered. Closer perhaps to Venturi on the other hand is Alexander's language of iconic patterns, an environmental and structuralist 'pattern language'. This linguistic approach to form is identified in Alexander's case with a reverence to innate human associations with formal types. Alexander contends that designers should draw on these natural icons rather than seek to continually reinvent new forms solely for the sake of individual style or novelty. His notion of a pattern language thus tracts right beside the messy vitality that Jacobs and Venturi use to separate the authentic inclusive wholes from the contrived authoritarian ideals.

In Alexander's words: 'We are searching for some kind of harmony between two intangibles: a form which we have not yet designed and a context which we cannot properly describe' (1964: 26–27). Or to rhyme back to the wicked encounter with complexity: 'Complexity is one of the great problems in environmental design. Adequate information about the existing environment and about the types of place that it is desirable to make cannot be kept inside one brain' (Alexander, 1966).

Collectively, these three positions on messy architectural complexity demand an authentic and meaningful fit within an inclusive whole. Fit here refers both to the cohesion of component pieces and to the setting of a completed solution within its context. In closing this messy encounter, it is worth noting that all three of the champions, Venturi, Jacobs and Alexander, wrote from within the mainstream of architecture and urban design.

ENCOUNTER THREE: ORDERED COMPLEXITY OF THE ESSENCE

Different views and approaches to the question, 'what is going to count as information?' can lead to radically different problem spaces and certain kinds of view can ensure that the problem space used is not an adequate map of the task environment. It is necessary, therefore, to examine at some length the relation between knowledge and architecture (Heath, 1984: 25).

Ordered complexity can thus be described as the process of reeling in and organizing all the project information designers need to digest, even as building programmes are getting more and more complicated. Sanoff (1968) and Peña (1969) along with Preiser (1978) pioneered this, and many practitioners of programming and strategic planning today are their protégés. Work in this area of architectural complexity is principally concerned with finding the unique essence of a design situation so that its true nature can be addressed. Ordered complexity deals with the rational tasks of programming and strategic goal setting, but is also interested in collaborative problem definition, participation of multiple stakeholders and in proactive scenario planning. The latter is what we associate with Buckminster Fuller as 'inventing the future' as opposed to passively following current trends. In this proactive sense, there is an architectural connection between ordered complexity and wicked complexity. At the same time, Venturi's complexity overlaps with the problem definition and goal-setting agenda of ordered complexity:

> *First the medium of architecture must be re-examined if the increased scope of our architecture as well as the complexity of its goals are to be expressed. Simplified or superficially complex forms will not work ... Second, the growing complexities of our functional problems must be acknowledged. (Venturi, 1966: 19)*

Many architects choose to subjugate this ordered complexity aspect of their work into a category of 'predesign' and view it as a marginal but necessary distraction from their core pursuit of design as an art form. Peña (1969) actually advocates a separation between problem seeking and problem solving, precisely because of this prevalent alienation between the cultural artistic pursuits of architecture as distinct from its social and ethical missions. But this alienation is also a powerful dialectic which activates a more holistic approach to ordered complexity, because at the core of even the most rational mindsets in this literature of ordered complexity, there is a constant search for systemic solutions to what can only be called wickedly complex problems. Under the procedurally analytic approach of every major thinker in this area, this structuralist investigation is presented as the ultimate pursuit.

In other terms, ordered complexity is a quest for the unique essence of a design situation. As a teleological principle, this unique essence can be positively construed as Aristotle's Final Cause, where 'an acorn is potentially an oak tree'. Final Cause asks, for example, why the little acorn (much to the disappointment of hungry squirrels everywhere) does not simply grow into a larger acorn. Teleology illustrates that the acorn's process of change and motion is ultimately directed at becoming an oak tree. In complexity, as in architecture, we can understand the acorn as an instruction set of what is to be designed, but not how to go about formalizing it. Architecture's encounter with ordered complexity is the signal of just such a quest for essential design instruction.

As a notion of design paralleling that of chaos theory, ordered complexity can also be conceived of as how we identify the core problem space which organizes all the facts and figures collected in 'predesign'. Surrounding this core is a negative space that can be thought of as the region of fertile ambiguity. In systems theory, this ambiguity can be characterized as what occurs on the edge of chaos where organisms adapt to environmental feedback in a cybernetic loop of indeterminate actions and reactions. Without ambiguity and the hysteresis of feedback, the system falls to homeostasis and ceases to function. In the design space and task environment of architecture, it is this very ambiguity that makes design possible.

To restate hysteresis and homeostasis back into both complexity science and natural complexity in architecture, consider the work of British chemist James Lovelock (1979) who began in the 1960s to compare the atmosphere of Mars, which he observed was in a state of dead equilibrium, while the atmosphere of Earth was in a constant state of dynamic, self-correcting, self-evolving and self-regulating equilibrium. This led him to propose the Gaia Principle, which considers Earth as one immense interactive living system, every part of its complex workings affecting every other part.

In ordered complexity literature, the niche of ambiguity is occupied quite neatly by works dealing with scenario planning or, more poetically, with Fuller's admonition to 'invent the future'. Scenario planning and the field of strategic planning in general is fundamentally concerned with separating the problem space into core known determinates, peripheral unknowable noise and thereby revealing the region between them of ambiguity and choice (Figure 2). In architecture, this framework of thinking could liberate designers to focus on the nascent and pregnant possibilities where ambiguity truly exists, and where design resources are not wasted in futile efforts to reframe the fixities or ponder futile impossibilities. It is for reasons such as these that design tends to often become less creative by modifying existing solutions that offer robust precedents, as opposed to exploring radically different and potentially superior ones. To paraphrase the poets:

> *The primary [spontaneous] imagination I hold to be the living power and prime agent of all human perception ... The secondary [contemplative imagination] I consider as an echo of the former, coexisting with the conscious will, yet still identical with the primary in the kind of its agency, and differing only in degree, and in the mode of its operation [dissolve, diffuse, idealize, dissipate, unify] ... Fancy, on the contrary, has no other counters to play with, but [ready-made] fixities and definites. The fancy is indeed no other than a mode of memory. (Coleridge, 1817: 167)*

After unique-essence and rich-ambiguity, the third component of ordered complexity is a social one, that of multiple stakeholders with conflicting interests, varying priorities and potentially different definitions of the design mission altogether. These stakeholders may consist of different client or user groups or of different members of the project team. A client, for example, may involve directors of different departments as well as managers of facilities, finance, human resources and a host of other individuals with vested interests and rival responsibilities. Likewise, a hospital must work not just for doctors, but also for patients, nurses, technicians and visitors – not to mention the hospital authority. On the team side, major building projects are led by an architect who orchestrates a team of civil, structural and mechanical engineers, as well as a host of consultants. Where the client stakeholders

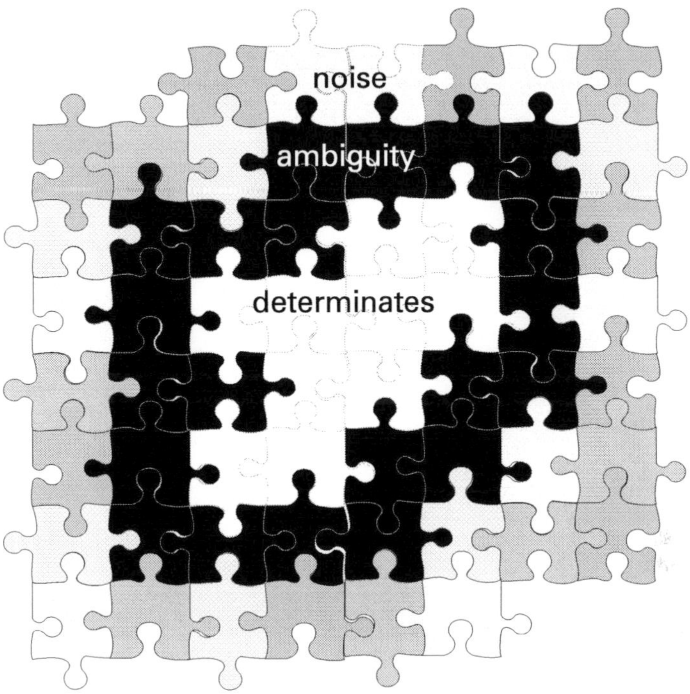

FIGURE 2 Diagram of determinates, noise and ambiguity

represent potentially antagonistic goals of the overall project, the design team stakeholders work on different components of the project with potentially antagonistic priorities and differing means of implementation.

Team building in architecture is captured in the discipline of project management. Here we have the procedural thinking that begins perhaps with Frederick Taylor and develops through mathematical techniques such as Program Evaluation and Review Technique (PERT) or Critical Path Method (CPM). The collaborative and collective intelligence involved in project management is expected to reflect professional attitudes and capabilities of each and every participant. It is not so surprising that an element of complexity would be assumed in this work. The client and user side of the conflicting stakeholder issue has not been so cleanly addressed however. While some works tackle the issue of user groups and multiple stakeholders directly (e.g. Sanoff, 1978, 2000), many architects are focused on the owner as the sole source of design parameters. These designers are primarily concerned with a normative translation of the owner's brief into a programme and agenda for design.

Despite this three-fold encounter with unique essence, scenario planning and collaborative design, the general history of architecture's encounter with ordered complexity does not yet adequately account for complex underpinnings. Nor does it grasp their sources of potential inspiration. Instead, designers seem to generally prefer a quick overview of the design parameters with just enough depth to arrive at an intuitive proposition about how to proceed. Hence, as mentioned earlier in this section, designers must ultimately repeat the same basic solution types that have been used before with only incremental and superficial adaptations. Connections back to Venturi's 'inclusive whole' and Jacobs 'authentic fit' are thus weakened by reliance on robust precedents. To quote Alexander again:

The crucial quality of shape, no matter of what kind, lies in its organization, and when we think of it in this way we call it form ...Today functional problems are becoming less simple all the time. But designers rarely confess their inability to solve them. Instead, when a designer does not understand a problem clearly enough to find the order it really calls for, he falls back on some arbitrarily chosen formal order. The problem, because of its complexity, remains unsolved. (Alexander, 1964: 1)

ENCOUNTER FOUR: NATURAL COMPLEXITY OF THE ORGANISM

A look behind the footlights reveals that nature has no choice in the assignment of roles to players. Her productions are shoestring operations, encumbered by the constraints of three-dimensional space, the necessary relations among the sizes of things, and an eccentric sense of frugality. (Stevens, 1974: 4)

Finally, we turn to architecture's encounter with natural complexity, presented here as the intentions of design to incorporate or emulate nature's own complex responses to design's specific challenges. As an architectural encounter, this complexity can be described as having five dimensions – ecology, flow, morphogenesis, synergistics and Gestalt psychology.

The ecological dimension of natural complexity begins in architecture with Patrick Geddes (1904, 1911), a Scottish biologist and pioneering urban ecologist who initially worked with Thomas Huxley. Geddes is also closely associated with the work of his protégé, the prolific historian Lewis Mumford (1922, 1924, 1939, 1947, 1967). For more details on the Geddes and Mumford connection, see Novak (1995). Together, Geddes and Mumford describe civilization in successive phases of paleotechnic, technic and then neotechnic eras of design; foretelling Bell's observation of our current post-industrial era transition into a society of knowledge-based production. Their neotechnic era also portends the ascendance of information and cybernetics over machine production and pollution.

Further elaboration of the ecological branch of this encounter emerged most prominently with Victor Olgyay's *Design with Climate* (1963) and Ian McHarg's *Design with Nature* (1969). These works came on the heels of Rachel Carson's *Silent Spring* (1962) and signal not only an emergence of green design imperatives, but also a corresponding set of design tools. Other primary works that were formative to this aspect of design thinking would include Givoni (1969) and Knowles (1969, 1974, 1981, 2006), then slightly later Mazria (Mazria, 1979; Mazria *et al*, 2007) and Lyle (1985, 1994). None of this thinking has tarnished in the years since publication and all the volumes mentioned are still foundational reading.

This ecological approach is principally one of seeking harmony with the patterns and forces of nature. It recognizes environment as context, treats climate as a resource and inspiration, and it seeks to discover form that is self-emergent and responsive rather than trying to wilfully invent it. In terms of architecture's encounter with complexity, the ecological approach treats built artefacts as if they were themselves an organism of nature.

Very close to this ecological theme is the dimension of natural complexity as organized flow. Steven Groák (1992), for example, imagined that a building may be considered as a system of flows – people, light, air, heat, information, products, gravity, sound and so on. The building's components then assume metaphysical description as reservoirs, conduits, capacitors and, to use Norberg-Schulz's (1966) original terms, filters, barriers and switches.

This perspective of buildings as sets of organized flow is important in how it animates the architectural concept of systems, as opposed to the conventional way in which systems are usually trivialized. Instead of a structural system of columns, beams and joists, for example, we can think about a system for how gravity courses through the building. Instead of corridors, rooms and doorways in an egress system, we have an organized set of flows for how people move through buildings.

Flow also connects back to the ecological theme. Lyle (1994) laid out 12 strategies relating design and flow. His most relevant points were to use intelligence to replace energy, allow multiple pathways for complex flows, use form to embody process, and shape form to accommodate flow.

Finally, flow also inspires morphological principles – the appropriation of biological order into artificial objects. Christopher Alexander argued this in his *Notes on the Synthesis of Form* (1964), a work for which he was awarded the American Institute of Architects' Gold Medal. Janine Benyus's *Biomimicry* (1997) is perhaps the most acclaimed current work in this theme. In general, these works are derived from the self-organizing and self-emergent form we find in nature as it is now understood to be the purpose-driven result of shaping and regulation by systems of organized flow. From snowflakes to skeletons to the stars, in the organized convection flow of boiling water, the choreography of a flock of birds, or the rippled crests of sand dunes, pattern and structure in nature always arise to embody their formative flows.

The next dimension of natural complexity, after those of ecology, flow and morphogenesis relates directly to the difference between the whole and the sum of its parts. Buckminster Fuller (1963, 1969, 1981; Fuller and Applewhite, 1975) popularized this as the 'synergistic' quality of how the behaviour of systems differs from what can be expected by examining their parts as separate elements. To Fuller, synergy was more than a principle, it was a cause leading to the 'ephemeralization' of a problem by doing more and more with synergistic design that required fewer and fewer resources. His own architectural projects, such as the Dymaxion House (1948) and his dome for the American Pavilion at Montreal's Expo '67, illustrated this through what he labelled 'tensegrity', the building of strong structures from thin triangulated members. As a connective note to wicked complexity, tensegrity is also used in management science as a concept in deliberative structures, for which Beer (1994) developed a 'syntegration' method for looking at collaborative deliberation.

The synergistic dimension of natural complexity is also encountered in Bohm (1980) as a coherence of 'implicate and explicate order' where we find 'undivided wholeness in flowing movement'. On the less esoteric side is anthropologist and cyberneticist Gregory Bateson (1972, 1979) and his abductive or propositional view of examining patterns and relationships. More towards architecture, but still not purely design focused is Orr (1979, 1992, 1994, 2002). In mainstream architectural literature, synergy is often discussed in terms of integrated building systems (Guise, 1985; Rush, 1986; Bovill, 1991; Bachman, 2003).

The final dimension of natural complexity is found in the branch of Gestalt psychology that focuses on art and architecture, most prominent in the work on aesthetics by György Kepes (1944, 1965, 1966) and Rudolf Arnheim (1954, 1969, 1971, 1977). This theory of mind holds that the human perception of order recognizes holistic and self-organizing tendencies as well as dynamic interrelationships. Kepes was the director of the Center for Advanced Visual Studies at MIT, where he sought to unify nature and technology. As a longtime friend of Laszlo Moholy-Nagy, Kepes was an influential participant in the Bauhaus school. Arnheim had a similar agenda for unifying nature and technology and is best known for his landmark book, *Visual Thinking* (1969). To quote:

> *Order makes it possible to focus on what is alike and what is different, what belongs together and what is segregated. When nothing superfluous is included and nothing indispensable left out, one can understand the interrelation of the whole and its parts, as well as the hierarchic scale of importance and power by which some structural features are dominant, other subordinate. (Arnheim, 1971: 1)*

Or,

> *The aim of science we now see, is to find the relations which give order to the raw material, the shapes and structures into which the measurements fit ... the relations which the facts have with one another – the whole they form and fill, not with their parts. In place of the arithmetic of nature, we now look for her geometry: the architecture of nature. (Bronowski, 1965: 56)*

In summary, architecture's four encounters with natural complexity provide phenomenological and experiential linkages between built form and its complex operations. This natural complexity is manifested by ecological responsiveness to nature, ordering of flows, morphogenesis, synergistic integration, and human cognition of meaning through order. Natural complexity is also architecture's closest encounter with scientific complexity because of the direct connection of built form with environmental performance and the technically based methods of achieving such design intentions. As such, natural complexity comes ready with more instrumental design tools than do the encounters related to defining problem space – wicked problems and ordered essence. Likewise, natural complexity is more instrumental than the rich, messy complexity of Jacobs, Venturi and Alexander.

Despite this instrumentality and affinity with science, natural complexity is also the most ennobled of the four encounters. Current professional design standards have, for example, strongly embraced the ethic and the engaged pursuit of sustainable design. It remains to be seen, however, if scientific understandings of the building as a complex system will be fully acknowledged. At this point, architects are still developing the tools and agility to incorporate basic procedural advances in this area.

DISCUSSION

Summary illustrations are offered here to clarify the main points and chronologies as discussed in the four encounters. Table 1 lays out a knowledge map of each mode of architectural complexity as well as a corresponding column on the scientific view. Figure 3 sketches the major aspects of each encounter and identifies their primary advocates.

The motivation for sketching out this missing history originates with the recognition that architectural design has had significant encounters with complexity, but that these events have never been drawn together or given coherence. Such coherence will only be sufficient when designers recognize how far their skirmishes with complexity have ventured into realms of complexity science and how far there is yet to go. Perhaps as the two histories continue to converge, architecture will accept that these are shared realms and will commit to its own program of wicked, messy, ordered and natural design investigations. This is not to say that architecture and science are merging or that one is minor to the other. But it can be argued that both disciplines operate within the same parameters of a complex and dynamic reality. We may thus expect that they have some common basis in complexity.

In closing, there are two points of difference between architectural complexity and complexity science to clarify. The first is the difference between systems and symptoms: in the era of post-industrial knowledge production and mounting accountabilities such as sustainability, post-occupancy evaluation and continuous commissioning, architecture will be

TABLE 1 Complexity science and four modes of complexity as encountered by architecture

	SCIENTIFIC	WICKED	MESSY	ORDERED	NATURAL
Proponents	Ackoff, Arnhof, Churchman. . .	Simon, Rittel, Bell. . .	Jacobs, Venturi, Alexander	Pena, Sanoff, Preiser. . .	Geddes, Kepes, Olgyay. . .
Post-war landmarks	1957	1957	1961	1965	1963
Realm	Cosmology	Society	Culture	Institution	Organisms
Objective	Adaptation	Intelligence	Spontaneity	Essence	Responsive
Organization	Dynamic	Cybernetic	Organic	Collaborative	Holistic
Order	Interactive	Managed	Authentic	Discovered	Emergent
Problem	Evolve	Bounded rationality	Identity	Reduce data	Flow
Agents	Systems	Decision makers	Citizens	Stakeholders	Systems
Application	Behaviour	Organization	Urban	Definition	Design

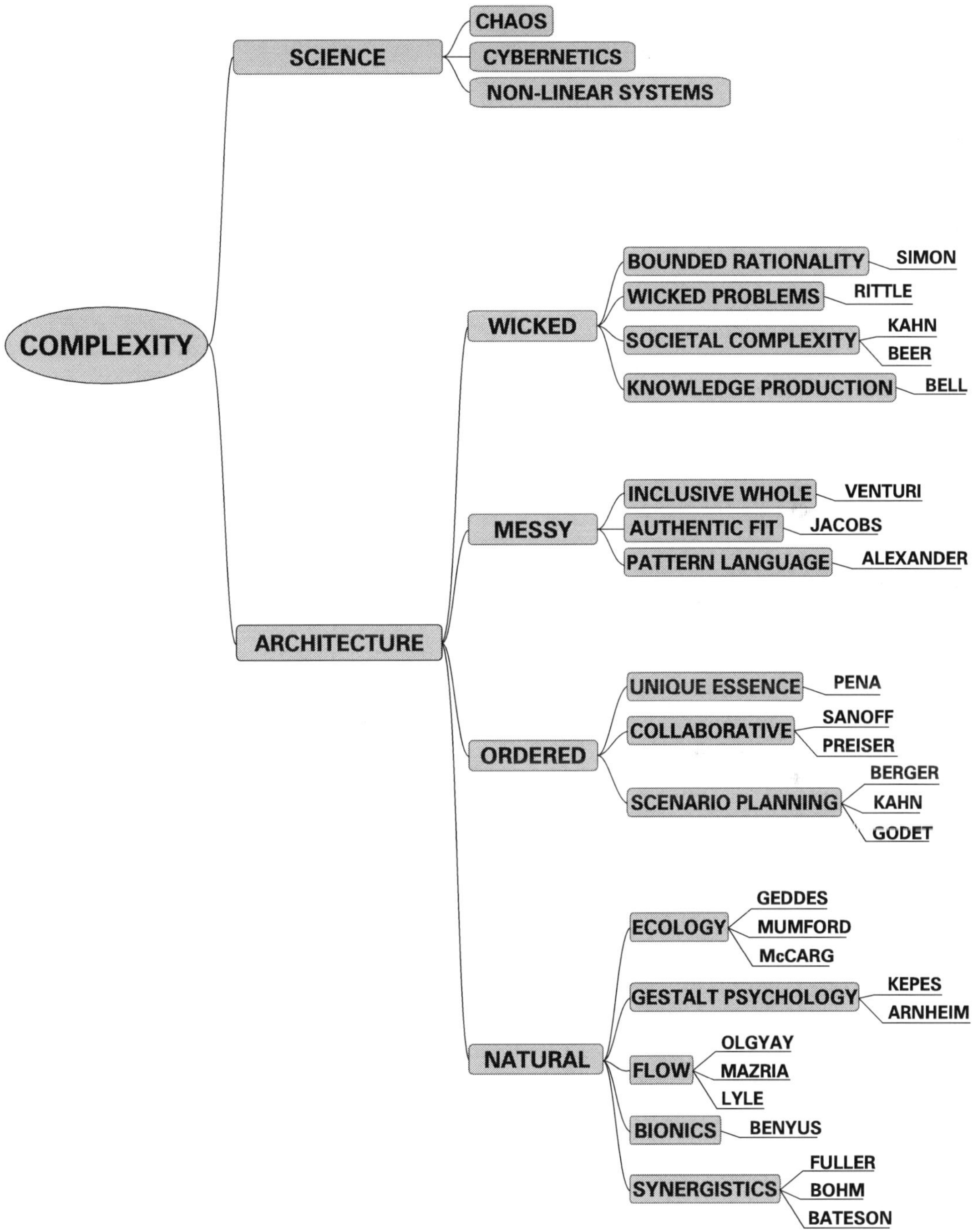

FIGURE 3 Knowledge map of architecture's four encounters with complexity (composed with FreeMind, available at http://freemind.sourceforge.net/)

challenged to provide systemic solutions to increasingly wicked problems in a world of diminishing robust solutions. Formal gestures will not suffice. The second point is that of scale and representation: complex systems operate not just in the local scale of human perception, but also in the non-local scales of micro and macro forces far outside direct human experience. The architecture of this non-local world is clearly not amenable to the sort of representation that architects associate with their most cherished productions. A different vision is required.

The gift that architecture brings us, however, is an endless faith and capacity for the translation of relations into abstractions, abstractions into ideas, and of ideas into forms, places, situations and cities. It is to this faith that architecture must turn when it finally encounters complexity on its own wicked terms.

> *We must face the fact that we are on the brink of times when man may be able to magnify his intellectual and inventive capacity; just as in the nineteenth century he used machines to magnify his physical capacity. Again, as then, our innocence is lost. And again, of course, the innocence, once lost, cannot be regained. The loss demands attention, not denial. (Alexander, 1964: 11)*

AUTHOR CONTACT DETAILS

Leonard R. Bachman: University of Houston, 122 College of Architecture, Houston, Texas 77204-4431, US. Tel: +1 713 743 2372, fax: +1 713 743 2358, e-mail: LBachman@UH.edu

REFERENCES

Ackoff, R.L., 1968, *Fundamentals of Operations Research*, New York, Wiley.

Ackoff, R.L., 1974, *Redesigning the Future: A Systems Approach to Societal Problems*, New York, Wiley.

Alexander, C., 1964, *Notes on the Synthesis of Form*, Cambridge, Harvard University Press.

Alexander, C., 1966, *The City As a Mechanism for Sustaining Human Contact*, University of California, Berkeley, Center for Planning and Development Research.

Alexander, C., Ishikawa, S. and Silverstein, M., 1977, *A Pattern Language: Towns, Buildings, Construction*, New York, Oxford University Press.

Arnheim, R., 1954, *Art and Visual Perception: a Psychology of the Creative Eye*, Berkeley, University of California Press.

Arnheim, R., 1969, *Visual Thinking*, Berkeley, University of California Press.

Arnheim, R., 1971, *Entropy and Art: an Essay on Disorder and Order*, Berkeley, University of California Press.

Arnheim, R., 1977, *The Dynamics of Architectural Form: Based on the 1975 Mary Duke Biddle Lectures at the Cooper Union*, Berkeley, University of California Press.

Bachman, L.R., 2003, *Integrated Buildings: The Systems Basis of Architecture*, Hoboken, NJ, John Wiley & Sons.

Bateson, G., 1972, *Steps to an Ecology of Mind: Collected Essays in Anthropology, Psychiatry, Evolution, and Epistemology*, San Francisco, Chandler Publishing Co.

Bateson, G., 1979, *Mind and Nature: A Necessary Unity*, New York, Dutton.

Beer, S., 1994, *Beyond Dispute: The Invention of Team Syntegrity*, New York, Wiley.

Bell, D., 1973, *The Coming of Post-industrial Society: a Venture in Social Forecasting*, New York, Basic Books.

Benyus, J.M., 1997, *Biomimicry: Innovation Inspired by Nature*, New York, Morrow.

Bohm, D.,1980, *Wholeness and the Implicate Order*, London, Routledge and Kegan Paul.

Bovill, C., 1991, *Architectural Design: Integration of Structural and Environmental Systems*, New York, Van Nostrand Reinhold.

Bronowski, J., 1965, 'The discovery of form', in G. Kepes (ed), *Structure in Art and in Science*, New York, G. Braziller.

Buchanan, R., 1995, *Discovering Design: Explorations in Design Studies*, Chicago, University of Chicago Press.

Carson, R., 1962, *Silent Spring*, Boston, Houghton Mifflin.

Churchman, C.W., 1967, 'Wicked problems', in *Management Science*, 14(4), Application Series B141–B142.

Coleridge, S.T., 1817, *Biographia Literaria: or, Biographical Sketches of my Literary Life and Opinions*, New York, C. Wiley & Co, Chapter XIII, 167.

Fuller, R.B., 1963, *Ideas and Integrities, a Spontaneous Autobiographical Disclosure*, Englewood Cliffs, NJ, Prentice-Hall.

Fuller, R.B., 1969, *Utopia or Oblivion: The Prospects for Humanity*, Toronto, New York, Bantam Books.

Fuller, R.B., 1981, *Critical Path*, New York, St Martin's Press.

Fuller, R.B. and Applewhite, E.J., 1975, *Synergetics; Explorations in the Geometry of Thinking*, New York, Macmillan.

Fuller, R.B. and Walker, E.A., 1970, *Approaching the Benign Environment*, Auburn, Alabama, published for Auburn University by University of Alabama Press.

Geddes, P., 1904, *City Development, a Study of Parks, Gardens and Culture-institutes: A Report to the Carnegie Dunfermline Trust*, Edinburgh, Geddes and Co.

Geddes, P., 1911, *Evolution*, New York, H. Holt and Co.

Geddes, P., 1950, *Cities in Evolution*, New York, Oxford University Press.

Givoni, B., 1969, *Man, Climate, and Architecture*, New York, Elsevier.

Groák, S., 1992, *The Idea of Building: Thought and Action in the Design and Production of Buildings*, London and New York, E&FN Spon.

Guise, D., 1985, *Design and Technology in Architecture*, New York, Wiley.

Heath, T., 1984, *Method in Architecture*, Chichester, West Sussex and New York, Wiley.

Horn, R.E., 2001, *Knowledge Mapping for Complex Social Messes*, a presentation to the Foundations in the Knowledge Economy at the David and Lucile Packard Foundation, 16 July, http://www.stanford.edu/~rhorn/a/recent/spchKnwldgPACKARD.pdf on (accessed 13 March 2008).

Jacobs, J., 1961, *The Death and Life of Great American Cities*, New York, Random House.

Jacobs, J., 1969, *The Economy of Cities*, New York, Random House.

Jencks, C., 1995, *The Architecture of the Jumping Universe: a Polemic: How Complexity Science is Changing Architecture and Culture*, London, Academy Editions.

Kepes, G., 1944, *Language of Vision*, Chicago, P. Theobald.

Kepes, G., 1965, *Structure in Art and in Science*, New York, G. Braziller.

Kepes, G., 1966, *The Man-made Object*, New York, G. Braziller.

Knowles, R., 1969, *Owens Valley Study: a Natural Ecological Framework for Settlement*, Los Angeles, National Endowment for the Arts.

Knowles, R., 1974, *Energy and Form: an Ecological Approach to Urban Growth*, Cambridge, Mass, MIT Press.

Knowles, R., 1981, *Sun Rhythm Form*, Cambridge, Mass, MIT Press.

Knowles, R., 2006, *Ritual House: Drawing on Nature's Rhythms for Architecture and Urban Design*, Washington DC, Island Press.

Laurence, P., 2006, 'Contradictions and complexities', in *Journal of Architectural Education*, 59(3), 49–60.

Lovelock, J., 1979, *Gaia, a New Look at Life on Earth*, Oxford and New York, Oxford University Press.

Lyle, J.T., 1985, *Design for Human Ecosystems: Landscape, Land Use, and Natural Resources*, New York, Van Nostrand Reinhold.

Lyle, J.T., 1994, *Regenerative Design for Sustainable Development*, New York, John Wiley.

Mazria, E., 1979, *The Passive Solar Energy Book: A Complete Guide to Passive Solar Home, Greenhouse, and Building Design*, Emmaus, PA, Rodale Press.

Mazria, E., Fisk, P. III, Reynolds, J. and Roaf, S., 2007, *The 2030 Challenge: Environmental Design in the Face of Climate Change*, Fall 2006 Klai Juba Lecture Series, Las Vegas, UNLV School of Architecture, http://www. library.unlv.edu/arch/events/2006fall2030.html (accessed 13 March 2008).

McHarg, I., 1969, *Design with Nature*, Garden City, NY, Natural History Press.

Mumford, L., 1922, *The Story of Utopias*, New York, Boni and Liveright.

Mumford, L., 1924, *Sticks and Stones: A Study of American Architecture and Civilization*, New York, Boni and Liveright.

Mumford, L., 1939, *Men Must Act*, New York, Harcourt Brace and Co.

Mumford, L.,1947, *Green Memories: The Story of Geddes & Mumford*, New York, Harcourt Brace and Co.

Mumford, L., 1967, *The Myth of the Machine: Technics and Human Development*, New York, Harcourt Brace Jovanovich.

Norberg-Schulz, C., 1966, *Intentions in Architecture*, Cambridge, Mass, MIT Press.

Novak, F.G., 1995, *Lewis Mumford and Patrick Geddes: The Correspondence*, London and New York, Routledge.

Olgyay, V., 1963, *Design with Climate: Bioclimatic Approach to Architectural Regionalism*, Princeton, NJ, Princeton University Press.

Orr, D.W., 1979, *The Global Predicament: Ecological Perspectives on World Order*, Chapel Hill, University of North Carolina Press.

Orr, D.W., 1992, *Ecological Literacy: Education and the Transition to a Postmodern World*, Albany, State University of New York Press.

Orr, D.W., 1994, *Earth in Mind: On Education, Environment, and the Human Prospect*, Washington DC, Island Press.

Orr, D.W., 2002, *The Nature of Design: Ecology, Culture, and Human Intention*, New York, Oxford University Press.

Peña, W.M., 1969, *Problem Seeking: New Directions in Architectural Programming*, Houston, Caudill Rowlett Scott.

Preiser, W.F.E., 1978, *Facility Programming: Methods and Applications*, Stroudsburg, PA, Dowden, Hutchinson & Ross.

Rittel, H. and Webber, M., 1973, 'Dilemmas in a general theory of planning', in *Policy Sciences*, 4, 155–169.

Rush, R.D. (ed), 1986, *The Building Systems Integration Handbook*, New York, Wiley.

Sanoff, H., 1968, *Techniques of Evaluation for Designers*, North Carolina State University, Raleigh, Design Research Laboratory, School of Design.

Sanoff, H., 1978, *Designing with Community Participation*, Stroudsburg, PA, Dowden, Hutchinson & Ross.

Sanoff, H., 2000, *Community Participation Methods in Design and Planning*, New York, Wiley.

Simon, H.A., 1957, *Models of Man: Social and Rational; Mathematical Essays on Rational Human Behavior in Society Setting*, New York, Wiley.

Simon, H.A., 1969, *The Sciences of the Artificial*, Cambridge, Mass, MIT Press.

Simon, H.A., 1976, *Administrative Behavior: A Study of Decision-making Processes in Administrative Organization*, New York, Free Press.

Stevens, P.S., 1974, *Patterns in Nature*, Boston, Little Brown.

Van Emden, M.H., 1971, *An Analysis of Complexity*, Mathematical Centre Tracts, Volume 35, Amsterdam, Mathematical Centrum.

Venturi, R., 1966, *Complexity and Contradiction in Architecture*, Museum of Modern Art, distributed by Doubleday, Garden City, NY.

ARTICLE

The systemic nature of the construction industry

José L. Fernández-Solís

Abstract

This paper addresses the 'know why' we build by looking first at the systemic nature and complexity that informs the construction industry's current paradigm. An objective of this paper is to present how the construction industry's reigning paradigm is driven by an obfuscating manufacturing–industrial–engineering mindset. Under the NAICS (North American Industry Classification System of the US Census Bureau), the construction industry is listed in the service sector of the economy under section 23 and is broken down into many categories, such as Buildings (236) and Heavy and Civil (237). Construction is considered a 'basic industry', like manufacturing, mining, fishing and farming. This industrial classification system implies a scientific–industrial–manufacturing–managing paradigm that is expected to overarch all industries. Over the past 10 years, Latham (1994) and Egan (1998, 2002) have challenged the industry to improve its efficiency as well as the quality of its output. Failure by the industry to achieve these efficiencies points to a lack of understanding of the systemic nature of the industry. The author is interested in the 'why' that explains the 'how'. If construction cannot be viewed in the same light as a manufacturing industry, then what is the paradigm of the construction industry?

■ *Keywords* – Adaptive change; panarchy; paradigm; philosophy; systemic nature of the construction industry; theory

INTRODUCTION

Considerable effort in construction is dedicated to increasing efficiencies based on improving 'know how' through Lean Construction (LC), Critical Path Method (CPM), Last Planner System (LPS) and Just in Time (JIT) movements among others, experiments using different project delivery systems such as Design-Bid-Build (DBB), Design-Build (DB), Construction Manager at Risk (CM@r) to name a few, and the application of technological and informational systems.

The reigning paradigm dictates that construction should follow industrial trends of increased efficiency, control, quality, productivity and overall decrease in cost per unit as seen in the automobile, shipping and aerospace industries (Ranta, 1993). However, these efficiencies have not been achieved in construction. Why? There are several reasons to be considered: a simple answer may be that the part of construction moving off-site is the very efficient part (thus it is counted under manufacturing). So the construction industry is measured by all the on-site production, which is an intrinsically inefficient part. This shift has been going on for decades. Construction may be getting more efficient overall, but it never shows in the statistics of 'construction'.

For the argument in this paper, construction consists of diverse sub-industries supplying products and services to one-off projects. This uniqueness has allowed the architectural profession to promote expressiveness of built form as the great common good, above efficiency and repeatability. The latter is seen as having lower cultural value.

However, the author argues that perhaps the construction industry is radically different from manufacturing and thus the industrial paradigm is not

doi:10.3763/aedm.2008.S807 ISSN: 1745-2007 (print), 1752-7589 (online) www.earthscanjournals.com

the best fit. This paper addresses the need to 'know why' we build by looking first at the systemic nature and complexity that informs the construction industry's current paradigm.

INVOKING PHILOSOPHY IN CONSTRUCTION THINKING

We need to understand 'why' as well as 'how' construction occurs. The urgent problems connected with technology require philosophical clarification (Kuhn, 1962, 2000; Eriksson, 1997) since much that has been written about construction's paradigm is inadequate (Ballard and Howell, 2003), making it all the more important for serious philosophers to get involved.

Koskela and Kagioglou (2006) show how philosophy (until recently considered an obscure and antiquated field of knowledge and, according to some, superseded by science and technology) influences worldview (paradigm), which trickles down to science, technology, processes and products (Nightingale, 2000). Therefore, the author will address the construction paradigm issue, using logic, epistemology, metaphysics and/or philosophical thinking to establish foundational knowledge concerning our discipline.

TWO METAPHYSICAL WORLDVIEWS: ENTITIES AND BECOMING

The current worldview of an industry is based on increases in efficiencies that can be gradual, step or radical. Construction as an industry is expected to achieve the efficiencies of other industries, such as those building aircraft, ships and automobiles.

The anomalies between the desired efficiencies of construction, when compared with other industries and actual practice, have led to an in-depth examination of the industry (Koskela, 2000) beginning with its peculiarities, characteristics and attributes. These studies highlighted the types and amount of anomalies in construction as an industry, comparing them with other industries in terms of efficiencies, cost reduction, innovation and, lately, theory building.

Koskela (2000), when comparing construction with other industries, identified the peculiarities of construction to be:

- each project is generally one of a kind – a unique prototype
- each project is carried out in situ, exposed to the weather and with particular site conditions
- each project is carried out by an assembled team that may be different on each project.

Bridging the gap between construction and normal manufacturing industry brought forth the characteristics of transformation, flow and value as dominant concepts for management and control. Furthermore, extensive work has been done on the attributes of planning, dispatching and constructing (Ballard, 2000).

The gap between manufacturing and normal industrial (other than construction) efficiencies, as well as the lack of efficiencies in the construction industry, were attributed to a lack of theory in the industry. One of the reasons for this anomaly was advanced by Koskela and Vrijhoef (2001) who suggested that the construction industry lacks its own body of theories – a quest that has led the author to the First International Symposium on the Theory of the Built Environment 2007 in Salford, UK. The search for a theory for the construction industry may be closer to the metaphor alluded to in Gunderson and Holling's *Panarchy: Understanding Transformations in Human and Natural Systems* (2002), where a theory of adaptive change is proposed. In short, panarchy is a metaphor of how complex natural and human systems change, i.e. how they are transformed. The construction industry is a human system but it is the complex nature of the industry that makes it behave as an adaptive system, thus its transformation in time is captured by the panarchy metaphor – a subject for further detailed studies.

THE SYSTEMIC NATURE OF THE CONSTRUCTION INDUSTRY

Two main aspects of construction are of particular interest and in need of better definition: one relates to an understanding of the systemic nature of building construction as an 'industry' and the other to its 'complexity'. Furthermore, the author postulates that the intertwined, dynamic, complex characteristics of building construction (Nam and Tatum, 1988) is the arena where one can observe a paradoxical co-dependency of product (what is, being, entity, artefact) and process (what should be, becoming).

MODELS FROM OTHER INDUSTRIES

The models of the aircraft, shipbuilding, automobile and other industries should be sufficient to manage construction complexity, if building construction has the characteristics of these mass-production industries. However, if building construction does not reflect the characteristics of these industries, then a new model is needed.

For example, imagine how the aircraft industry would operate if it worked along the same lines as construction. Each artefact (airplane) would be constructed for one client with differing requirements from any other, for one specific use (say one route). Although the plane could be constructed out of an existing catalogue of parts, it would be designed and built by a team that came into being just for this purpose. Furthermore, the airplane would have to be built or assembled on a site open to the elements. The artefact would be a unique but distinct product, meeting governing requirements and obeying the laws of nature.

Academicians and some practitioners in the construction industry believe that theories and practices from one industry are translatable to the other. What prevents theoretical translation are differences in scale and differences in emphasis between product and processes.

Construction in general does not behave as an 'industry' but more like a 'conglomerate of industries', an 'industry of industries' or a 'meta-industry'. The meta-industry metaphor borrows Palmer's (2003, 2004) concept of meta-system which by definition includes holes, absurdities, inefficiencies and paradoxes as well as the capacity to invent and innovate. If this is the case, past behaviour of specific industries is not directly translatable to the behaviour of a meta-industry, mainly because of the subtle but real differences in scale and domains. This line of generic and structural thinking regarding complexity and the systemic nature of building construction as a meta-industry requires additional foundational work. The author may look, in future work, at Gunderson and Holling's *Panarchy* (2002) and the theories of adaptive change as a better fit for construction.

TOWARDS AN UNDERSTANDING OF THE SYSTEMIC NATURE OF THE INDUSTRY

Koen (1985, 2003) succinctly states that the engineering method under which building construction can be located is based on 'change', utilizing available resources and some 'particular rationality' (albeit heuristic rather than scientific in his view). This rationality is derived from 'the state of the art' at that point in time, directed towards a 'best or optimum solution', but always occurring in an 'environment of uncertainty'. All types of engineering, science and philosophy fall under the category of heuristics, according to Koen (2003). Koen proposes the following example of heuristic rationality:

> *At the appropriate point in a project, freeze the design; allocate resources as long as the cost of not knowing exceeds the cost of finding out; allocate sufficient resources to the weak link; and solve problems by successive approximation.*

In a construction project, uncertainties (Bertelsen, 2003, 2004, 2005; Bertelsen and Emmitt, 2005) are the result of temporary coalitions in a turbulent environment requiring semi-predictable or even unpredictable configurations of supply industries and technical skills. Groák (1992, 1994) and Polanyi (1967, 1974) call these 'technological paradigms' organized around a 'project' rather than the 'firm or production process' (Nightingale, 2000), which is a major paradoxical distinction between construction and manufacturing. The 'firm', as used by Nightingale, refers to service providers. Paradoxically, although the project as an axis or centre of attention is essential, the defining characteristic of the systemic nature of building construction is that it is a dynamic process; in short, the dynamic process is the centre of attention (see Figures 1 and 2).

A clear distinction between construction and the traditional definition of an industry such as manufacturing is essential for an understanding of the systemic nature of the 'industry'. The capacity of the industry (Hillebrandt, 1975; Pearce, 2006) or potential

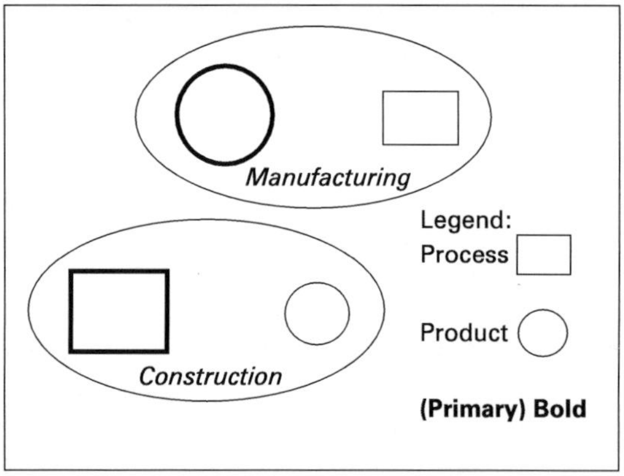

FIGURE 1 Systems abstraction

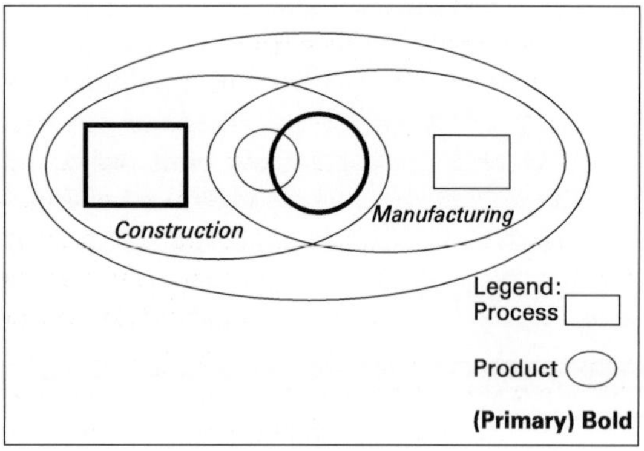

FIGURE 2 Building construction abstraction

is revealed in the use of different distinct resources and skill bases for different building types and different construction sectors, i.e. civil, building, industrial, manufacturing, housing and medical. These construction sectors are in fact different 'industries'; according to Kodama (1992): 'We are moving away from the idea of "one technology, one industry" as the framework of analysis for building construction capacity for change.' In addition, not only are we moving away from an understanding of construction as encompassing only one industry, but also an understanding of construction as encompassing several industries (Kodama, 1992) and, furthermore, moving away from the model of an 'industry' as understood in manufacturing and defined by 'industrial science' theories and practices. Is this a semantic or real distinction?

Groák (1992, 1994) states 'we should no longer treat construction activities as belonging to "an industry' with definable boundaries, specific technical skills and using specific resources'. The focus is continually shifting more and more towards its end-products and services, recognizing increasing external linkages and potential innovations from beyond 'construction' where the construction capacity resides according to Hillebrandt (1974, 1975, 1984), a position embraced by Pearce (2003, 2006) and followers.

METAPHYSICAL BASIS FOR DISTINCTIONS BETWEEN PRODUCT AND PROCESSES

Koskela and Kagioglou's (2006) research states that since the pre-Socratic period, there have been two basic metaphysical worldviews. The thing-oriented view seems to lead to analytical decomposition, the requirement or assumption of certainty and a more static philosophical approach. They argue that 'production' is intrinsically a process-oriented endeavour. However, an analysis of current conceptualizations shows that it is the thing-oriented view of the world (product) that has dominated research and practice of production management (Nightingale, 2000). What the authors mean by this is that research and production management practices have used the Cartesian method of problem decomposition (Descartes' (1898) second rule in Cottingham (1986)). Thus, according to Koskela and Kagioglou (2006), the general direction of research (and production management) is achieved by going into even smaller parts of the whole and searching for explanations at the lowest possible reductionist level, as used by Sir Isaac Newton and followers, also known as the scientific approach.

The two underlying assumptions behind the thing-oriented worldview as related to decomposition are first, similarity and, second, independence of decomposed elements or parts. Koskela and Kagioglou (2006) state:

> *The similarity assumption takes it for granted that the parts are, by nature, similar to the whole and thus also mutually similar. The assumption*

of the independence of parts follows from the similarity assumptions. Namely, if our unit of analysis is an idea, problem or thing in itself, so will all decomposed parts also be ideas, problems or things in themselves.

On the other hand, process metaphysics holds that 'everything flows' and is in 'change' (Weick, 2000). According to Rescher (2000), a contemporary understanding of process metaphysics, as quoted by Koskela and Kagioglou (2006), follows:

- time and change are among the principal categories of metaphysical understanding
- processes are more fundamental than things (i.e. projects) for the purposes of ontological theory
- contingency, emergence, novelty and creativity are fundamental categories of (process) metaphysics.

Rescher (2000) defines process as a structured sequence of successive stages or phases, having three characteristics (thus establishing the criteria for processes):

- a process and a product are duals of each other
- a process and a product have a relativistic space–time and time–space relation to each other found in relativity theory
- there is a process phase and a product phase of the phase space of the entity
- processes and products vary along a spectrum from simple to complex
- both processes and products can have parts, but the parts of processes are staged in time
- processes take place in relation to a reference while products appear in the background of other processes
- thus perceptually, processes and products are based on flow and gestalt. Flows can be seen as temporal gestalts, where it takes time for the figure to appear on the background, while spatially simultaneous to the appearance of products on their ground
- unity and totality and wholeness of processes and products are not intrinsic but something that appears as we look at the mereology (whole part relations within the process and product)
- at the pattern level, there are structures and flows as well as signs and values
- at the form level there is the shape and behaviour of an object as well as states and events (such as receiving messages).

Construction, meaning to build, as a verb or activity, is about change. To have a building is to have first the activity that created the building, as understood by Aristotle. Imai's (1986) *Kaizen* (a Japanese model for continually increasing quality) recognizes two types of change – abrupt change such as the difference between two sets of things (i.e. the natural and the artificial environments) and change as the process between the now and the after-now.

Changes occur at a macro level (industry, economy, society) and micro level (the firm, project-specific organization and the project itself). The essence of this activity is environmental change – where there was nothing, now there is a building, through the process of construction. Because the arena of change is the natural environment interacting with an artificial environment, it can be argued that building construction, as well as construction in general, is a process, but with a project's outcome (product) as its essential secondary axis. As a process, it is always 'now this, then that'; it is complex as we have noted, with a temporal and non-eliminable spatial–temporal dimension; furthermore, the building construction process has its own structure.

In contrast, manufacturing is based on a product as its main axis of attention, but with a process as its essential secondary axis (see Figure 1). This is like the difference between the space–time and the time–space matrix. Manufacturing operates in space–time and construction in time–space (Heidegger, 1962).

Mass product manufacturing is a tightly coupled system (with product as its main axis and process as the enabling characteristic), whereas building construction is a loosely coupled system (Dubois and Gadde, 2002), with process as its primary axis of attention and product as its secondary axis or enabling characteristic. This distinction between manufacturing and construction highlights the difference between the two systems (Nightingale, 2000). Mass product manufacturing, as a tightly coupled industrial system, exhibits the following characteristics:

- delays are not allowed or possible
- sequence of events are invariant (except in flexible manufacturing schemes)
- alternative paths are tightly controlled or not available
- little or no opportunity for substitution or repair (usually discarded, wasted)
- slack is not desirable
- redundancies are designed and deliberate.

In contrast, building construction, as a loosely coupled system, exhibits the following characteristics (Nam and Tatum, 1988; Dubois and Gadde, 2002):

- number of permutations and possible combinations are enormous (Weick 1994, 2000)
- complex operations (Baccarini, 1996; Gidado, 1996)
- inefficient operations (Cox and Townsend, 1998)
- sub-optimization (Gann, 1996)
- some tightly coupled, some time-sensitive specialized activities with sequentially interdependent activities with standard parts (Gidado, 1996)
- mostly, it is a loosely coupled system (Dubois and Gadde, 2002)
- overlapping activities – long lead time and slack built in
- adaptive on-site changes (Vrijhoef and Koskela, 2005) and consequential changes (Crichton, 1996)
- generation of variations (Akintoye *et al*, 2000)
- self-determination – coordination with different firms, each adding a measure of slack
- work is redone when non-conforming, rather than product discarded as in manufacturing.

It is reasonable to infer that building construction as a process is bounded at the upper end of the taxonomy by systems (and even meta-systems) with complex process-driven entities. However, at the same time, the boundaries at the lower end are assemblies that are product-driven entities (see Figure 2). Perhaps this duality of process and product underlies the thinking of the proponents that want to make building construction more like manufacturing, i.e. product driven.

Building construction's capacity for change, its potential, is therefore an intrinsic source as well as a recipient of variability, inefficiency, non-linearity and comfort, with chaos, creativity, novelty, uniqueness and even paradox and ambiguity, i.e. all the problems associated with embodiment. A high capacity for change implies freedom at many levels of the panarchy. In panarchy, there is an arena where systems are nested and go through a process of adaptive change. Construction exhibits a parallel with the panarchy metaphor in that it is an arena where systems are nested and go through adaptive change in the process of realising their potential. In other words, the meta-systems nesting allows a high degree of inventiveness, promotes creativity, and celebrates diversity and adaptive change. However, like in panarchy, as the potential is reached, the systems' coupling becomes tighter and increasingly loses the freedom, inventiveness, diversity and ability to change.

The capacity for change is furthermore exacerbated at the product end and the client himself is a complex system and source of variability (Cherns and Bryant, 1984; Pries *et al*, 2004). This 'product' axis of the paradox also exhibits the characteristics of uniqueness, expression, one-of-a-kind, on a particular site with particular characteristics, actors selected and acting autonomously (Koskela, 2000).

It is then prudent to say that from both the supply and the demand side, from the process and the product, and as a matter of fact from the milieu where building construction takes place, the fundamental characteristics of building construction are those of a complex system, process driven with a normative capacity for change.

THE COMPLEX NATURE OF THE INDUSTRY

According to Chu *et al* (2003), there is no generally accepted definition of complexity, no universal and unified Theory of Complexity (TOC) and, according to critics, complex systems are too diverse to share any profound 'common causes for common characteristics'. Furthermore, Chu has also observed that at a higher level in the field of the philosophy of science, there is no unique, simple criterion or litmus test to decide if a theory is scientific or not. Thus rather than looking for a universal criterion for being scientific, it is often better to ground criteria in the *aim* of the theory or a heuristic (Koen, 2003). According to Chu *et al* (2003), three aims are central:

- Predictive component – prediction of the future behaviour of a system, given a set of observational data about it; an active quantitative prediction and experimental manipulation of phenomena.
- Exploratory component – theoretical understanding and/or explanatory description of a system/framework for a number of phenomena.
- Control component – provision of guidelines and control mechanisms for the intervention and manipulation of a system; ability to manipulate the exploitation of scientific theory.

Ideally then, a scientific theory would explain, predict and facilitate control at the same time. However, all scientific theories do not follow the list of mentioned components but may emphasize one of the components while the others are treated as negligible factors.

Chu *et al* (2003), in addition to the above criteria, make the following pertinent observations (emphasis added):

> *A central and related issue is the language in which a TOC is to be formulated. Science is largely dominated by a Platonist ideal (Koyre, 1968). The essence of this ideal was established in mechanics by Galileo and its most important success is … theoretical physics. Often a TOC is (more or less tacitly) assumed to be a mathematical theory (Gödel, 1931). Holland, for example points out that the mathematical form has the additional advantage of high precision and generalization. One may add that prospects of prediction and control might look better if a mathematical form is possible. Indeed quantum theory is wholly formalized and quantitative. In contrast, the theory of biological evolution by means of natural selection, for example, involves mathematics only for the formulation of detail, whereas the main insight is formulated in natural language. A TOC might be of this latter kind.*

Borrowing from Koen (2003), TOC may be realistically considered 'heuristic'. An additional important property of TOC is universality – a TOC should be applicable to most, if not all, of a wide range of different complex systems. Chu *et al* (2003) state that:

> *In a Platonist/Galilean science tradition the idea of 'universal theories' is often equated with 'unified theories'. Although unification is of a highly aesthetic value, it should not be regarded as a litmus test for a universal TOC.*

The author would expect a TOC to be useful in controlling natural systems or to be predictive or to be explanatory. It should make some claim of universality. However, one would expect a possible trade-off between universality and mathematical quantitativeness and it may or may not exhibit 'unification' characteristics.

SYSTEMS THINKING AND THE CONSTRUCTION INDUSTRY

Richmond (1993) uses the term 'systems thinking' as a replacement for the term 'systems dynamics'. Richmond suggests seven critical 'systems thinking' skills and the need to operate on all seven thinking tracks simultaneously. The term 'systems thinking' is somewhat of a paradox (Zemke, 2001; Davidz *et al*, 2006), since this phrase combines words that imply individual and multi-actor concepts into one research construct. This creates difficulty in selecting the unit and level of analysis. The unit of analysis is the entity being described or analysed during a research study (in construction the unit and level of analysis can be macro, i.e. industry and micro, i.e. project and firm). Four standard levels of analysis in social sciences are individuals (inventors, entrepreneurs), groups (social and cultural groups), organizations (building construction) and environments (natural and artificial environments). Individual characteristics, group dynamics, organizational culture and surrounding environments all affect the system of interest in multi-level interactions, thus producing a 'wicked' problem atmosphere.

Richmond (1993) suggests seven critical systems thinking skills – dynamic thinking, closed-loop thinking, generic thinking, structural thinking, operational thinking, continuum thinking and scientific thinking. He suggests that good 'systems thinking' *tracks simultaneously all seven skills*. Once the unit and level of analysis are established, it is possible to make some observations on the nature of the building industry.

OBSERVATIONS ON THE NATURE OF THE BUILDING INDUSTRY

Complexity in building construction (Bertelsen, 2005) has been studied from different perspectives; for example, as part of managing complexity in project production (Bertelsen and Koskela, 2002, 2003) and seeing the client as a complex system (Pries and Janssen, 1995; Lucas, 2000, 2004, 2005; Bertelsen, 2003; Emmitt and Gorse, 2003; Pries *et al*, 2004; Bertelsen and Emmitt, 2005).

Managing complexity in project production is an attempt (Shewhart, 1931; Shewhart and Deming, 1939) to minimize variability to get the productive activity under control. The aim of this approach is to avoid complexity and uncertainty, which could disturb tight controls. Recently this approach has been championed in Lean Construction (Alarcón, 1997; Ballard *et al*, 2002; Koskela *et al*, 2002), Last Planner (both are service trademarks of the Lean Construction Institute, US) (Ballard, 2000) and Just in Time theories and practices. However, as noted by Bertelsen (2005), there are production situations with inherent complexity and unpredictability that escape efforts at reducing complexity, codifying procedures, learning to improvise, and buffering.

These approaches result in a paradoxical inverse linkage of effectiveness and efficiency. Efforts at reducing complexity also rub against the nature of a meta-system characterized as a fertile breeding ground for inventions and innovations, where technologies created for other sectors are adopted and assimilated successfully. In other words, control of waste is the opposite of variability, and variability is the breeding ground of possibilities. Hence, building construction has unrecognizable and unrecognized complexity (as in a 'meta-system') and acts more like a complex, dynamic, living organism that is self-organizing (Kauffman, 1993, 1995) and adaptive as in a learning organization (Edmondson and Moingeon, 1998) rich with waste and with innovation possibilities. There are complex systems and complex meta-systems. These two are different – this subject will be further elaborated in future work.

Client complexity (Pries *et al*, 2004), and its attendant consequences, showcase chaos theories (Lorenz, 1972, 1993; Waldrop, 1992) where the lack of initial definition or moving targets along the phases create ripe conditions for divergence. This approach is characterized by non-linearity and richness in feedback loops where a formal analytic approach is no longer possible or desirable. A new way of looking at complex systems is provided by Lucas (2000), who identifies a comprehensive list of 18 characteristics found in complex systems. This list is considered by other researchers (Bertelsen, 2005) as fairly exhaustive. A closer examination of these characteristics as related to building construction, allows a grouping into three categories – autonomous agents, undefined values, and non-linearity. Furthermore, Bertelsen (2005) relates these types to one of the three aspects of construction advanced by Koskela (2000) – transformation, value and flow.

However, what is considered a holistic systems view is also considered a reductionist view when the boundaries of the point of perspective are redrawn. Two examples are proposed, one from the aircraft industry, and the other from the construction industry. Davidz *et al* (2006) supplies the following example that has been adapted: Consider an aircraft engine – a 'system' could be a part (a set of compressor blades called the compressor stage), a component (a compressor), a sub-system (an aircraft engine), a production system (an aircraft), a group of design engineers (Advanced Compressor Design), a business (GE Aircraft Engineers), a larger business (GE), a sector (the aerospace sector), a national general economy (US), or the global system (air transportation). The definition of a system and the point of view that makes it holistic is therefore driven by the end-state or application of interest.

Building construction, in contrast, has a component (motor), a production system (mechanical system), a business (air-conditioning), a larger business (a building) with multiple and disparate systems (elevator, plumbing, electrical, structural) that make it a meta-system, which is part of the building sector, part of construction, a national general economy and the global system of satisfying human needs for security and shelter from the natural environment.

Building construction fits as follows (see Figure 3). Building construction (as part of construction in general, not shown for clarity) is listed as a subset of the industrial sector of the general economy. The industrial sector also forms the background where the system's components, sub-systems, suppliers, vendors and manufacturers reside. Hence, building construction, a dynamic process that we have seen, is bracketed by the notion or reality implied in the word 'industry'. Bertelsen (2005) groups 14 of the complex system characteristics (taken from Nam and Tatum, 1988; Lucas, 2004) into the three previously noted categories. The author has taken this list, expanded it to the full 18 elements (see Table 1) from Lucas (2000, 2004, 2005) and added information from Koskela (2000) and Shingo (1988). The following expands Lucas' definition of the elements found in construction complexity:

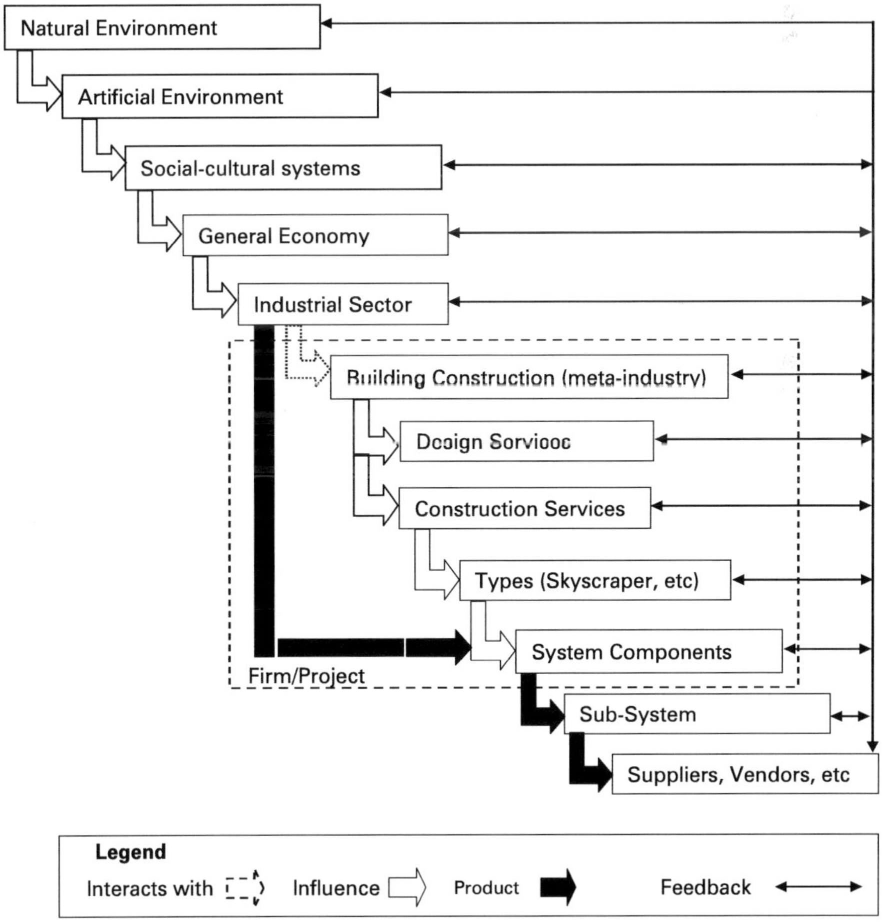

FIGURE 3 Proposed general taxonomy

TABLE 1 Complex systems

KOSKELA: TRANSFORMATION	VALUE	FLOW
SHINGO: OPERATIONS	VALUE	PROCESS
1.1 Autonomous agents	2.1 Undefined values	3.1 Non-linearity
1.2 Non standard	2.2 Fitness landscape	3.2 Emergence
1.3 Co-evolution	2.3 Non-uniform	3.3 Attractors
1.4 Self-modification		3.4 Phase changes
1.5 Downward causation		3.5 Unpredictability
1.6 Self-reproduction		3.6 Instability (variability)
1.7 Mutability		3.7 Learning organization
1.8 Fuzzy functions		

1.1 Autonomous agents – Stakeholders are varied, not identical, with differing perspectives and interests, which change over time.
1.2 Non-standard – The system is heterogeneous and allows varying associations over time.
1.3 Co-evolution (self-organization) – The parts may evolve in conjunction with each other in order to fit into a wider system.
1.4 Self-modification – Parts can change their associations or connectivity freely.
1.5 Downward causation – A system is made up of its parts, and the parts are affected by the emergent properties of the whole system.
1.6 Self-reproduction – The system can replicate itself.
1.7 Mutability – Random interval changes may occur in the system.
1.8 Fuzzy functions – The overall function (purpose) of the system is co-evolved.
2.1 Undefined values – The meaning of the system's interface with its environment is not specified at the outset.
2.2 Fitness – The distribution of choices can be modelled using the concept of fitness landscapes, with local optima and global optimum that are relative and dynamic.
2.3 Non-uniform – The system is different and evolves in time in response to internal and external demands.
3.1 Non-linearity (non-equilibrium) – The system operates far from equilibrium since it takes energy from its environment.
3.2 Emergence – System properties are higher-level meta-systemic functions of the system (Peitgen, 1986); emergent phenomena travel from specifics to generalities and vice versa as well as from systems to ambience and vice versa.
3.3 Attractors – The system has multiple dynamic attractors; it can be stable for a while, but not permanently.
3.4 Phase changes – The feedback may lead to sudden jumps to another relatively stable phase.
3.5 Unpredictability – The system is chaotically sensitive to its initial conditions.
3.6 Instability – Over the long-term, step changes or catastrophes occur.
3.7 Learning organization – The organization evolves by learning from experience and errors.

The state-of-the-art research in peer-reviewed publications derives these complex characteristics from the construction industry. Lucas' 18 characteristics as grouped by Bertelsen (2005) (and adapted in this analysis) with the headings provided by Koskela (2000) and Shingo (1988) are then analysed using as a background Richmond's seven skills for complex thinking. Complexity and flexibility are the 'coins' in the realm of systems that allows the expending of complexity 'dollars' to achieve useful gains such as increased functionality, efficiency and/or flexibility. For example, we now consider the three main groups identified by Bertelsen (2005) – autonomous agents, undefined values and non-linearity – as they apply to building construction, using a systems thinking approach.

Autonomous agents

Complex systems, such as building construction, are composed of independent or autonomous agents that are not identical. By design, no permanent executive or directing node exists in the system, with a control structure or leadership that emerges through self-organization. This self-organization (Kauffman, 1993) evolves and acts as an emergent control structure for the product and the process (Abdelhamid, 2004). These characteristics highlight non-linearity, learning organization (Edmondson and Moingeon, 1998), self-organization, and downward process causation, with high variability and therefore unpredictability.

The autonomous agent element is found between levels of the schema (subcontractors, system components, construction services, design services), as well as within each level (i.e. construction services). For example, in construction services, one autonomous group (the formal control structure) plans the project with a management-as-planning-and-dispatch modality and another autonomous group (the informal control structure) builds the project where self-modification, learning organization, and dealing with unpredictability and variability establish non-uniform variation from the formal control structure (Koskela and Howell, 2002).

A design is uniquely created for the one-of-a-kind, on-site project by a team assembled for the task (varying from project to project) which necessitates a co-evolution of the process, the organization and the project that is unique to the site and the owner's need. This co-evolution of elements aims to fit the created 'fitness landscape' and achieve the 'local maxima' (terms explained in the next section) while keeping in mind the global maximum that has been (sometimes) established (to varying degrees), but that can change during the project's gestation. New materials, methods, design solution, processes and others require self-modification and a learning organization.

Undefined values

The boundary of a system in the building construction arena is not initially specified and evolves in dynamic communication to fit a landscape that, itself, is emerging. The concept of fitness landscape in this sense reflects the finding of the 'local optima' high ground, for each part of the project, the process and the organization (be it design or construction or both). This is advanced through mutations with regard to an established (directly, indirectly or implied) relative global optimum in a balancing act between quality, cost and time as well as efficiency and effectiveness. However, this process does not occur at one level but at multi-levels in the schema, creating a matrix environment of negotiations with no global optimum possible, a major characteristic of a complex system and thus a wicked problem.

The undefined value alludes to the non-standard and non-uniform one-of-a-kind product and process that happens in space and time. The freedom of association or movement permits the clumping and de-clumping over time as the self-organization structures itself, the process as well as the product. This dynamic freedom aims towards the goal of creating value, albeit undefined, as we shall see. This value exists even if a project ends up in court, as it generates value for the litigation system at the expense of other stakeholders.

Every project establishes (explicitly or implicitly and to varying degrees) its own economic fitness landscape, quality fitness landscape, performance fitness landscape and cost fitness landscape during the initial design stage. Fitness landscape in this sense can be considered as a term that encompasses, among others, the concepts underlying building performance indicators.

'It is the nature of the project that it exists in its own fitness landscape,' states Bertelsen (2005). This fitness landscape includes, from the organizational side, the design and production (construction) components in differing possible combinations called project delivery systems. However, the building construction fitness landscape is emergent upstream with the industrial sector, the general economy and social system, among others; and emergent downstream with subcontractors, specific manufacturers and vendors, to name a few. Each possible combination eventually affects the project's local optima (component of phase) as well as the project's global optimum. Therefore, local optima, due to the aforementioned unpredictability, attractors, phase changes and variability (instability), create local optima and a global optimum much lower than expectations.

Unpredictable local optima has been the major complaint (Fisher, 1993; Gann, 1996; Johnson, 1995; Winch, 2003) when comparing building construction with other sectors such as manufacturing, where aircraft (Barber *et al*, 1998; Voodijk and Vrijhoef, 2003), ships and large complex systems reside but with differing parameters and control structures. The drive behind the move towards a manufacturing-based construction industry is a desire for increased efficiencies, i.e. reducing initial cost or labour, as in other industries (Latham, 1988, 1994; Egan, 1998, 2002). In Latham's and Egan's opinion, the construction industry is fragmented, a flaw to be remedied (Gann, 1996; Pyke, 2002; Woudhuysen and Abley, 2004).

However, Dubois and Gadde (2002) state that recent researchers, besides viewing industry outputs and outcomes as important, are valuing (in addition to costing) a diverse, open and flexible industry as something to be celebrated. The key in these contrasting positions is that manufacturing, as it is presently configured, is for mass production and this introduces a rigidity and inflexibility that runs contrary to the design and construction industry in its current configuration.

Bertelsen (2005) observes, and we concur, that there 'does not exist an absolute optimum' for a project, thus the 'best solution' is dependent on the system's real-world status. Others may counter that, although relative in time, with the state of the art, a performance-based building design coupled with a total building commissioning plan (NIBS, 1999), can establish a 'relative best solution' or 'relative optimum' that can serve as a project goal (Altwies and McIntosh, 2001). However, it is precisely problems without absolute optimal solutions that are wicked problems since the pre-conditions change as the solution evolves (Weick, 2000).

The project represents an emerging physical structure, and emerging process, through an emerging organization in time and space with increasing order and purpose and the intent of generating a specific value. This value, although constructed in the narrow sense for the client, ends up being a value for all levels of the general schema and thus the longevity and open structure of the industry, where variability is a key complexity characteristic.

Non-linearity

Complex systems are non-linear: the outputs are not proportional to the inputs and the whole is different from the sum of its parts. A roof is just a roof, but a house without a roof has much less value, if any.

The project itself acts as a dynamic attractor for an emerging organization of otherwise disparate stakeholders – autonomous agents with alliance to their parent organization. However, the temporary project organization for space and time self-organizes in formal and informal structures as a service, production and process system. The project dynamic contains multiple and sometimes conflicting attractions creating a matrix of possible and different behaviours within the project duration and between projects by the 'varying team' factor. Within project 'partnering' is an attempt to bring in line the sometimes conflicting attractions (see Black *et al*, 2000 for success factors and benefits).

The initial configuration of the team is in constant flux, thus, a multitude of actors create the system history that, coupled with phase changes and milestones where transitions occur, not only places the project at the 'edge of chaos' but is also a difficult history to capture fully for lessons-learnt exercises that would benefit a learning organization. These transitions are critical points in connectivity requiring the system (project, process and organization) to self-modify, self-organize and co-evolve to manage the fitness landscape, local optima and global optimum. The project organization's limited time existence does not allow it to emerge to a higher plateau, even if it is an intermediary step to the next full level of meta-organization, such as a virtual organization, or a fully vertically and horizontally integrated organization as found in heavy industry, e.g. petro-chemical.

In heavy industry, such as in a refinery, the high risk level and complexity require a smaller team that is relatively constant due to long project duration and teams moving from one project to the next seamlessly. One-of-a-kind solutions are discouraged, and lessons learnt are fully captured (or catastrophes induce heavy insurance premiums that become a motivator to capture lessons learnt and apply them through standards, regulatory frameworks and codes).

In buildings, understood as an assembly of systems (a system of systems), the wicked nature of the design, as well as the construction processes and the client itself as a complex system (Emmitt and Gorse, 2003; Pries *et al*, 2004; Bertelsen and Emmitt, 2005), makes it possible and most probable that small differences between stakeholders will lead to vastly different solutions, processes and project organizations. In other words, chaotic sensitivity to initial conditions makes a project rich in unforeseen events, deviation from plans and variability, thus affirming the concept of contextuality of a complex system.

The most critical decisions by stakeholders are affected by the disposition or attitude towards what eventually becomes the project's driving principle – cost, value or quality – which changes throughout the project phases and duration beyond the initial construction, i.e. during its life cycle. The expressed, implied or inferred attitudes of the stakeholders, individually and collectively, introduce sensitivity to chaos that is magnified through the project. The non-linearity of the project carries through the non-linearity of the building construction subsector, the non-linearity of the industry and the general economy. These additional levels of non-linearity upstream and downstream create a complex dynamic mechanism that is almost impossible to study in holistic or particular detail.

Lastly, building construction is nested in a social system with a varying team where communications and cooperation are emergent phenomena in each project. Cooperation is based on the project attractor for common behaviour, sometimes based on experience and performance, but often based on a low bid. Common behaviour stabilizes the project's cooperation (Bertelsen, 2005) 'in either a good or bad way' creating the 'project culture'. This culture, however, reflects all the characteristics and attributes of the larger social culture that includes cooperation, fighting or fleeing behaviour along with a myriad of other social types of behaviour. Thus culture adds another dimension of unpredictability, with a number of entrants and departures disturbing the 'project culture' stability in completely unforeseen ways.

This analysis of complexity in building construction does not pretend to be exhaustive. However, it illustrates how the project, process, emergent organization, culture and the other mentioned characteristics of complexity, such as autonomous agents, undefined value and non-linearity, are intertwined. We postulate that the dichotomy of project/process in building construction is so blurred that it may be theorized that a building is a project whose emergent and intrinsic quality is that of a process where the project, along with organization and all the supporting networks of services and products, forms the fundamental elements of the process. In Palmer's (2003, 2004) terminology, it is a meta-system formed by a paradox with two axes, project and process, and everything that comes within the boundary of that paradox will inherit the intrinsic paradoxical qualities. The project defines existence, while the process defines movement, if a differentiation needs to be made, although in reality they are one because they are mutually co-dependent – another characteristic of the paradox in meta-thinking.

CONCLUSIONS

The author has analysed past and current studies as a point of departure for an enquiry into the nature of systems that could help identify the necessary and sufficient characteristics that allow a glimpse of the systemic nature of the construction industry.

Construction's reigning paradigm of being an 'industry' does not quite capture the complexities embedded in the process of creating infrastructure and/or a building. The dynamics of construction go beyond systems dynamics due to the nature of its complexities. The application of current efforts to improve efficiency and effectiveness continue to generate significant differences between expectations and results. These differences point to anomalies between the current paradigm and its reality.

If the concept or proto-theory that views construction as a meta-industry has merit, the author conceptually discards 'industry' as the locus for finding solutions and opens up the possibility of looking elsewhere. A theory of construction should incorporate in a clear framework the workings of Lean, Building Information Modelling and Green initiatives in a framework akin to 'adaptive change'.

AUTHOR CONTACT DETAILS

José L. Fernández-Solís: Texas A&M University, 3137 TAMU, College Station, TX 77843-3137, US. Tel: +1 979 458 1058, fax: +1 979 861 1572, e-mail: jsolis@tamu.edu

REFERENCES

Abdelhamid, T., 2004, 'The self-destruction and renewal of lean construction theory: a prediction from Boyd's Theory', in *Proceedings of the 12th Annual Conference of the International Group for Lean Construction (IGLC12), Helsingore, Denmark*.

Akintoye, A., McIntosh, G. and Fitzgerald, E., 2000, 'A survey of supply chain collaboration and management in the UK construction industry', in *European Journal of Purchasing and Supply Management*, 6 (3–4), 159–169.

Alarcón, L., 1997, *Lean Construction,* Rotterdam, A.A. Balkema.

Altwies, J. and McIntosh, I.B.D., 2001, 'Quantifying the cost benefits of commissioning', in *Proceedings of the 9th National Conference on Building Commissioning, Cherry Hill, NJ, May 2001*, 1–13.

Baccarini, D., 1996, 'The concept of project complexity – a review', in *International Journal of Project Management*, 14(4), 201–204.

Ballard, G., 2000, *The Last Planner System of Production Control,* PhD thesis, Faculty of Engineering, University of Birmingham.

Ballard, G. and Howell, G., 2003, 'Competing construction management paradigms', in *Proceedings of the 2003 ASCE Construction Research Congress, Honolulu, Hawaii, March 2003*.

Ballard, G., Tommelein, I., Koskela, L. and Howell, G., 2002, 'Lean construction tools and techniques', in R. Best and G. de Valence (eds), *Design and Construction, Building in Value*, Oxford, Butterworth-Heinemann, 227–255.

Barber, P.R., Sheath, D.M., Walker, S., Graves, A.P. and Tomkins, C.R., 1998, 'A comparison of design management techniques in construction and the automotive industry', in Amor (ed), *Product and Process Modelling in the Building Industry*, Watford, Building Research Establishment, 67–74.

Bertelsen, S., 2003, 'Construction as a complex system', in *Proceedings of the 11th Annual Conference of the International Group for Lean Construction (IGLC11), Blacksburg, Virginia*, 11–23.

Bertelsen, S., 2004, 'Construction management in a complexity perspective', in *Proceedings of the 1st International SCRI Symposium, University of Salford, Salford.*

Bertelsen, S., 2005, 'Modularization – a new approach in making construction lean?', in *Proceedings of the 13th Annual Conference of the International Group for Lean Construction (IGLC13), Sydney, Australia.*

Bertelsen, S. and Emmitt, S., 2005, 'The client as a complex and chaotic system', in *Proceedings of the 13th Annual Conference of the International Group for Lean Construction (IGLC13), Sydney, Australia*.

Bertelsen, S. and Koskela, L., 2002, 'Managing the three aspects of production in construction', in *Proceedings of the 11th Annual Conference of the International Group for Lean Construction (IGLC11), Blacksburg, Virginia*, 13–22.

Bertelsen, S., and Koskela, L., 2003, 'Avoiding and managing chaos in projects', in *Proceedings of the 11th Annual Conference of the International Group for Lean Construction (IGLC11), Blacksburg, Virginia*.

Black, C., Akintoye, A. and Fitzgerald, E., 2000, 'An analysis of success factors and benefits of partnering in construction', in *International Journal of Project Management*, 18, 423–434.

Cherns, A.B. and Bryant, D.T., 1984, 'Studying the client's role in construction management', in *Construction Management and Economics*, 2(2), 177–184.

Chu, D., Strand, R. and Fjelland, R., 2003, 'Theories of complexity – common denominators of complex systems', in *Wiley Periodicals Inc.*, 8(3), 19–30.

Cottingham, J. (trans), 1986, *Rene Descartes in his Meditations on First Philosophy – With Selections from the Objections and Replies*, New York, Cambridge University Press.

Cox, A. and Townsend, M., 1998, *Strategic Procurement in Construction, Towards Better Practice in the Management of Construction Supply Chains,* London, Thomas Telford.

Crichton, C., 1996, *Interdependence and Uncertainty: A Study of the Building Industry,* London, Tavistock.

Davidz, H.L., Nightingale, D.J. and Rhodes, D.H., 2006, *Enablers, Barriers and Precursors to Systems Thinking Development: The Urgent Need for More Information,* Cambridge, Mass., MIT.

Descartes, R., 1898, *Discourse on Method and the Meditations,* translated by John Veitch, Amherst, NY, Prometheus Books.

Dubois, A. and Gadde, L.-E., 2002, 'The construction industry as a loosely coupled system: implications for productivity and innovation', in *Construction Management and Economics*, 20(7), 621–631.

Edmondson, A. and Moingeon, B., 1998, 'From organizational learning to the learning organisation', in *Management Learning*, 29(1), 5–20.

Egan, J., 1998, *Rethinking Construction: The Report of the Construction Task Force,* London, Department of the Environment, Transport and the Regions.

Egan, J., 2002, *Accelerating Change*, Construction Industry Council, http://www.strategicforum.org.uk/ (accessed 25 May 2006).

Emmitt, S. and Gorse, C., 2003, *Construction Communication,* Oxford, Blackwell Publishing.

Eriksson, L.J., 1997, 'How technology evolves', in *Complexity*, John Wiley & Sons, 2(3), 23–30.

Fisher, D., 1993, *Construction as a Manufacturing Process?* BAA Professor Inaugural Lecture, 18 May, University of Reading, Department of Construction Management and Engineering.

Gann, D.M., 1996, 'Construction as a manufacturing process? Similarities and differences between industrialized housing and car production in Japan', in *Construction Management and Economics*, 14(5), 437–450.

Gidado, K.I., 1996, 'Project complexity: the focal point of construction production planning', in *Construction Management and Economics*, 14(3), 213–225.

Groák, S., 1992, *The Idea of Building*, London, E&FN Spon.

Groák, S., 1994, 'Is construction an industry? Notes towards a greater analytical emphasis on external linkages', in *Construction Management and Economics*, 12(4), 287–293.

Gunderson, L.H. and Holling, C.S., 2002, *Panarchy – Understanding Transformations in Human and Natural Systems*, Washington, DC, Island Press.

Heidegger, M., 1962, *Being and Time,* New York, Harper & Row.

Hillebrandt, P.M., 1974, *Economic Theory and the Construction Industry,* London, McMillan.

Hillebrandt, P.M., 1975, 'The capacity of the industry', in D.A. Turin (ed), *Aspects of the Economics of Construction*, London, George Godwin, 225–257.

Hillebrandt, P.M., 1984, *Analysis of the British Construction Industry,* London, McMillan.

Imai, M., 1986, *Kaizen, The Key to Japan's Competitive Success,* New York, Random House.

Johnson, R.B., 1995, 'Making manufacturing practices tacit: a case study of computer aided production management and lean production', in *Journal of Operational Research Society*, 46, 1174–1183.

Kauffman, S.A., 1993, *The Origins of Order, Self-Organization and Selection in Evolution,* New York, Oxford University Press.

Kauffman, S.A., 1995, *At Home in the Universe, the Search for the Laws of Self-Organization and Complexity,* New York, Oxford University Press.

Kodama, F., 1992, 'Technology fusion and the new R + D', in *Harvard Business Review*, July–August, 70–78.

Koen, B.V., 1985, *Definition of the Engineering Method,* Washington, DC, American Society for Engineering Education.

Koen, B.V., 2003, *Discussion of the Method: Conducting the Engineer's Approach to Problem Solving,* New York, Oxford University Press.

Koskela, L., 2000, *An Exploration towards a Production Theory and its Application to Construction,* VTT Publications 408, VTT, Espoo, Building Technology.

Koskela, L. and Howell, G., 2002, *The Theory of Project Management – Problem and Opportunity,* Working Paper, VTT Technical Research Centre of Finland & Lean Construction Institute.

Koskela, L., Howell, G. and Tommelein, I., 2002, 'The foundation of lean construction', in R. Best and G. De Valence (eds), *Design and Construction: Building in Value,* Oxford, Butterworth-Heinemann, 211–226.

Koskela, L. and Kagioglou, M., 2006, 'On the metaphysics of production', in *Proceedings of the 13th Annual Conference of the International Group for Lean Construction (IGLC13), Sydney, Australia.*

Koskela, L. and Vrijhoef, R., 2001, 'Is the current theory of construction a hindrance to innovation?', in *Building Research and Information*, 29(3), 197–207.

Kuhn, T., 1962, *The Structure of Scientific Revolutions,* Chicago, The University of Chicago Press.

Kuhn, T., 2000, 'The road since structure', in J. Conant and J. Haugeland (eds), *The Road since Structure: Philosphical Essays 1970-1993*, Chicago, The University of Chicago Press, 90–104.

Latham, M., 1994, *Constructing the Team: The Final Report of the Government/Industry Review of Procurement and Contractual Arrangements in the UK Construction Industry,* London, HMSO.

Latham, M., 1998, 'Procurement: the present and future trends', in *CIB Procurement – The Way Forward,* CIB Publication 217, CIB, 61–74.

Lorenz, E., 1972, 'Predictability: does the flap of a butterfly's wings in Brazil set off a tornado in Texas?', in *Proceedings of the 13th Conference of the American Association for the Advancement of Science, Washington DC.*

Lorenz, E., 1993, *The Essence of Chaos,* London, UCL Press.

Lucas, C., 2000, *The Philosophy of Complexity,* www.caresco.org/lucas/themes.htm (accessed 7 July 2005).

Lucas, C., 2004, *Quantifying Complexity Theory*, www.calresco.org/lucas/quantify.htm (accessed 7 July 2005).

Lucas, C., 2005, *The Philosophy of Complexity,* revised www.caresco.org/lucas/philos.htm (accessed 7 July 2005).

Nam, C.H. and Tatum, C.B., 1988, 'Major characteristics of constructed products and resulting limitations of construction technology', in *Construction Management and Economics*, 6, 133–148.

NIBS (National Institute of Building Sciences), 1999, *Total Building Commissioning: Whole Building Design Guide*, Washington, DC.

Nightingale, P., 2000, 'The product-process-organization relationship in complex development projects', in *Research Policy*, 29, 913–930.

Palmer, K.D., 2003, *General Schemas Theory*, http://archonic.net (accessed 20 September 2005).

Palmer, K.D., 2004, *Nondual Science,* http://nondual-science.net (accessed 20 September 2005).

Pearce, D.W., 2003, *The Social and Economic Value of Construction: The Construction Industry's Contribution to Sustainable Development (Pearce Report)*, New Construction Research and Innovation Strategy Panel (nCRISP), London, http://www.ncrisp.org.uk (accessed 24 May 2006).

Pearce, D.W., 2006, 'Is the construction industry sustainable? Definitions and reflections', in *Building Research & Information*, Taylor & Francis, 34(3), 201–207.

Peitgen, H.O., 1986, *The Beauty of Fractals, Images of Complex Dynamical Systems,* New York, Springer-Verlag.

Polanyi, M.I., 1967, *The Tacit Dimension*, New York, Anchor Books.

Polanyi, M.I., 1974, *Personal Knowledge: Towards a Post-Critical Philosophy,* Chicago, The University of Chicago Press.

Pries, F., Doree, A., Van der Veen, B. And Vrijhoef, R., 2004, 'The role of leader's paradigm in construction industry change', in *Construction Management and Economics*, 22, 7–10.

Pries, F. and Janssen, F., 1995, 'Innovation in the construction industry: the dominant role of the environment', in *Construction Management and Economics*, 13(1), 43–51.

Pyke, S.D., 2002, 'Construction coalitions and the evolving supply chain management paradox: progress through fragmentation', in *Proceedings of COBRA 2002, Nottingham, UK, January 2002,* Royal Institute of Chartered Surveyors (RICS).

Ranta, J., 1993, *On the Dynamics and Evolution of Production Paradigms*, SITRA 130, Helsinki.

Rescher, N., 2000, *Process Philosophy,* Pittsburgh, University of Pittsburgh Press.

Richmond, B., 1993, 'Systems thinking: critical thinking skills for the 1990s and beyond', in *Systems Dynamic Review*, 9(2), 113–133.

Shewhart, W.A., 1931, *Economic Control of Quality of Manufactured Product,* New York, Van Nostrand.

Shewhart, W. and Deming, W.E., 1939, *Statistical Method from the Viewpoint of Quality Control,* The Graduate School, The Department of Agriculture, Washington.

Shingo, S., 1988, *Non-stock Production,* Cambridge, Productivity Press.

Vrijhoef, R. and Koskela, L., 2005, 'Revisiting the three peculiarities of production in construction', in *Proceedings of the 13th Annual Conference of the International Group for Lean Construction (IGCL13), Sydney, Australia, July*, 19–27.

Voodijk, J.T. and Vrijhoef, R., 2003, 'Improving supply chain management in construction: what can be learned from the aerospace industry?' in D.J. Greenwood (ed), *Proceedings of the Annual ARCOM Conference, University of Brighton, Brighton*, 2(3), 837–846.

Waldrop, M.M., 1992, *Complexity, The Emerging Science at the Edge of Order and Chaos,* New York, Penguin Books.

Weick, K.E., 1994, 'Small wins: redefining the scale of social problems', in *American Psychologist*, 39(1), 40–49.

Weick, K.E., 2000, 'Emergent change as a universal in organizations', in M. Beer and N. Noria (eds), *Breaking the Code of Change*, Boston, MA, Harvard Business School Press, 223–239.

Winch, G.M., 2003, 'The construction firm and the construction project: a transaction cost approach', in *Construction Management and Economics*, 7(4), 331–345.

Woudhuysen, J. and Abley, I., 2004, *Why is Construction so Backward?,* Chichester, UK, Wiley-Academy.

Zemke, R., 2001, 'Systems thinking: looking at how systems really work can be enlightening – or a wake up call', in *Training*, 39(2), 40–46.

ARTICLE

Models and metaphors: complexity theory and through-life management in the built environment

John Rooke, Ella-Mae Molloy, Murray Sinclair, Lauri Koskela, Mohan Siriwardena, Mike Kagioglou and Carys Siemieniuch

Abstract

Complexity thinking may have both modelling and metaphorical applications in the through-life management of the built environment. These two distinct approaches are examined and compared. In the first instance, some of the sources of complexity in the design, construction and maintenance of the built environment are identified. The metaphorical use of complexity in management thinking and its application in the built environment are briefly examined. This is followed by an exploration of modelling techniques relevant to built environment concerns. Non-linear and complex mathematical techniques such as fuzzy logic, cellular automata and attractors, may be applicable to their analysis. Existing software tools are identified and examples of successful built environment applications of complexity modelling are given. Some issues that arise include the definition of phenomena in a mathematically usable way, the functionality of available software and the possibility of going beyond representational modelling. Further questions arising from the application of complexity thinking are discussed, including the possibilities for confusion that arise from the use of metaphor. The metaphor of a 'commentary machine' is suggested as a possible way forward and it is suggested that an appropriate linguistic analysis can in certain situations reduce perceived complexity.

■ *Keywords* – Complexity; metaphor; modelling; production theory; unique adequacy

INTRODUCTION

Companies in the construction industry have tended to manage the risks of operating in the built environment (BE) through risk transfer moderated by contracts. However, these strategies do not in themselves reduce risk and more recently there has been an increasing tendency to consider construction and its products more holistically, with trends such as partnering, including public–private partnerships and through-life management, coming to the fore. The increased onus on players to consider the consequences of their actions in a wider sphere leads inevitably to a greater consideration of the complexity entailed in the design, construction and maintenance of the built environment. Moreover, these trends imply that we think not simply in terms of built environment products, but of the services (such as education, healthcare and communications) that buildings and civil engineering projects support, in terms of a product–service paradigm in the built environment. This paper seeks to explore how complexity theory may help to resolve the problems of organization thus presented.

Complexity concepts can be operationalized as either modelling or metaphor. We outline some

doi:10.3763/aedm.2008.S707 ISSN: 1745-2007 (print), 1752-7589 (online) www.earthscanjournals.com

parameters for the complex system under consideration and examine the application of complexity theory to the study of business. In conclusion, we focus on a particular metaphor which indicates the possibility of specifying the details of complex settings.

PRODUCTION AND SERVICE IN THE BUILT ENVIRONMENT

It is undeniable that companies and public agencies working to commission, design, build, service and maintain the built environment are operating in situations of complexity, both with regard to their environments and their internal operations. Considerations that feature in decision-making processes will include *inter alia*:

- fluctuations in available finance and thus effective demand
- evolving procurement and contracting strategies
- pressures to reduce costs, while improving timeliness and quality
- pressure to maintain profitability
- a range of competing initiatives to improve company performance
- differing and sometimes conflicting professional and occupational standards and loyalties
- the salience of industry-wide standards
- the need to improve safety performance and environmental sustainability
- the need to balance technological and organizational innovation against familiar and reliable methods.

From the perspective of through-life management, a further set of problems is added:

- the predictability of future service and maintenance costs
- future demand for the services the BE feature supports
- changing techniques and technologies in the services supported by the BE feature
- changing concerns and priorities with regard to the natural environment
- the acquired individual character of particular BE elements as they gain significance as local landmarks, cultural icons or historical sites.

METAPHOR

Morgan (1986) asserts the importance of metaphor in understanding organizations. His claim that 'our theories and explanations of organizational life are based on metaphors' (1986: 12) may be an over-generalization. However, he is undeniably correct in identifying the metaphorical process, in which we 'understand one element of experience in terms of another' (1986: 13), as one that is important to theory making.

Several benefits of the metaphorical use of complexity thinking can be identified; for instance, the importance of connectivity and interdependence in complex systems can be used to stress the collective nature of organization. Mittleton-Kelly (2003) reiterates a common social science observation that an individual's contribution to a team is constrained by what the team will permit (Parsons, 1968; Hofstede, 2002). She also notes that the social ecosystem in which an organization exists consists of other organizations which both affect that organization and are affected by it. She uses the concept of exaptation to emphasize the importance of allowing an organization's members to explore the 'space of possibilities' that exists, thus encouraging lateral thinking (de Bono, 1971). Along with Gershenson and Heylighen (2005), she stresses the concepts of self-organization, feedback and attractors to describe organizational change. Thus, complexity thinking is taken primarily as a stimulus to more flexible management styles, reinforcing what has long been a trend in management studies (Peters and Waterman, 1982; Handy, 1994).

In applying such a metaphorical approach to the built environment industry, particular peculiarities may be revealed. The so-called 'fragmentation' of the industry has long been commented on (see for instance Groak, 1994). More recently, aspects of fragmentation have been addressed from a more complexity-oriented perspective. Chen *et al* (2007) note the difficulties presented for knowledge management in multi-organizational systems. Bresnen *et al* (2005) point out the dispersed nature of management in project organizations and the difficulties this creates in organizational change. They adopt situated knowledge and symbolic interaction approaches to enable them to come to grips with this phenomenon. Bertelsen and Emmitt (2005) adopt an explicit complexity approach, to view the client as a complex system.

In an examination of three complex projects, two of which involved work on major built environment features, Ivory and Alderman (2005) derive four major implications for project management. They observe that systemic contradictions are most usually expressed as resource failures. Drawing on complexity analyses that use multi-nodality and multi-centredness, they conclude that some measure of mapping and prediction of knock-on impacts can be achieved, but they also observe that to a considerable extent, management activity consists of developing reactive organizational fixes. Interventions from subordinate nodes in the supply chain can have both beneficial and negative effects.

Drawing on the Danish Byg.SoL initiative, Bertelsen (2004) suggests a panoramic view of project management in the built environment that draws on the complexity metaphor to suggest five different ways of looking at management – planning, organizing, team building, providing a service, and causing a nuisance.

MODELLING

The use of mathematical modelling techniques for simulating complex systems within the built environment is controversial (Richardson *et al*, 2000). Often, a prediction of the future is expected as an output from a mathematical equation. Guastello (1995) states that the 'prediction of future states is possible in the near term, but could decay rapidly as we try to extend further into the future'. Although we know that maths cannot predict the future, we are disappointed when it does not.

Non-linear and complex mathematical modelling techniques have been applied to supply chain and workforce management (Guastello, 1995) and cellular automata for the modelling of regions (Allen, 1997a). Techniques that may be useful include fuzzy logic, cellular automata, chaos theory, catastrophe theory and fractals – basic explanations of which can be found in many texts (von Neumann, 1966; Codd, 1968; Kosko, 1993; Guastello, 1995; Berry, 1996; Wolfram, 2002; Steeb, 2005). Complex techniques are rarely used in isolation (Steeb, 2005), but more often in combination with other modelling techniques. For example, non-linear dynamic systems may be linked to game theory to investigate long-range planning and reactive management in the construction trades (Guastello, 1995). Fuzzy logic helps to explain the shades of grey in between the black and white of many established techniques.

Some suggest that while complex systems need not follow traditional modelling techniques, complex phenomena do not always need complex theory to explain them; complexity can emerge from simple beginnings (Wolfram, 2002).

There is much literature on agent-based models, multi-agent systems and Complex Adaptive Systems theory (the latter term originating from the work of John Holland and others at the Santa Fe Institute) (Carroll and Burton, 2000; Dijkstra *et al*, 2001; Shen *et al*, 2001; Seese and Schlottmann, 2003; Payr and Trappl, 2004; Robertson, 2005; McCarthy *et al*, 2006). These can demonstrate complex non-linear interactions that allow the emergence of new unpredictable properties in the system, allowing flexibility and a network-based interaction representation of built environment systems. Many focus on 'software agents', ignoring the special characteristics of human agents, but may still be useful to model teams, processes and organizations, for example. Agent-based models such as Conway's Game of Life (Gardner, 1970) have been in widespread use for around a decade.

Agent-based software programs available on the Internet include StarLogo (http://education.mit.edu/starlogo/) and JADE (http://jade.tilab.com). StarLogo is a basic programmable modelling environment for exploring the workings of decentralized systems (there is also an extension StarLogo TNG). It is designed as an introduction for students and researchers and has a colourful and usable interface. JADE (Java Agent Development Framework) is a software framework to develop agent-based applications, but seems to assume that agent = software agent, limiting its application to organizational systems.

MATLAB ® (www.mathsworks.com) is a high-level technical computing language and interactive environment for algorithm development, data visualization, data analysis and numeric computation. The many add-ons (toolboxes) it contains allow for flexibility in analysis and simulation. At its basic level,

MATLAB allows for the development of agents and their environment, while the toolboxes provide a wider range of application and accuracy.

Increasingly complex construction projects mean that the problems of uncertainty and ambiguity have ever larger implications because of the difficulty in predicting the impact of unexpected changes. Particularly promising fields for this endeavour appear to be economics and geography (Arthur, 1990, 1995; Allen, 1997a, 1997b), both of which are important to built environment concerns. Consideration of different aspects of design, such as environment, economy and society, in addition to the traditionally considered characteristics, makes the problem of design more complex. It has been suggested in the built environment context that mathematical techniques, such as fuzzy logic (Boussabaine and Elhag, 1999; Vakili-Ardebili and Boussabaine, 2007), stochastic approaches (Kirkham and Boussabaine, 2005) or neural networks (Boussabaine, 1996) may be useful, or even essential, to assist decision-makers. It is perceived that such techniques may aid circumnavigation of the subjectivity in design decisions, especially in modern challenges, such as ecological/sustainability concerns (Vakili-Ardebili and Boussabaine, 2007) and that these tools are becoming more accessible to non-computer specialists. Areas where the techniques have been applied include an approach to forecasting the remaining life of hospital buildings and optimization of maintenance strategies (Kirkham and Boussabaine, 2005); minimizing production time, achieving optimum mark-up estimation under different bid situations, predicting changes in construction cost indexes, estimating excavation efficiency based on attributes of operation elements and decision-making about construction modularization (Boussabaine, 1996); and cash flow (Boussabaine and Elhag, 1999).

The application of mathematical tools outlined above has generally been found to be useful, but they are underused because construction managers do not understand the mathematical processes by which the results are achieved. It may be possible to input a set of data and generate a suggested configuration, but without an understanding of the mathematical operations involved, it is difficult to have confidence in the outputs.

Another starting point for mathematical modelling may be found in the use of flow theory (Koskela, 1992, 2000). The application of flow theory in the built environment has led to useful developments such as the Last Planner (Ballard and Howell, 1998; Ballard, 2000) and the ADePT tool (Austin *et al*, 1994). However, notwithstanding the successful application of both these approaches (Thomassen *et al*, 2003; Baldwin *et al*, 2008) they do not represent the application of either sophisticated mathematics or modelling techniques. The Last Planner system is a management system that introduces flexibility and ultimately reliability into the planning system by shifting the locus of control down the management hierarchy and closer to the actual activity of production (Ballard and Howell, 1997a, 1997b). ADePT utilizes a dependency structure matrix (Austin *et al*, 1994, 2000) to rationalize the ordering of design tasks. One of the central phenomena of interest to flow theorists is that of queues (both Last Planner and ADePT address problems of queuing, although in radically different ways). Although queuing theory has been applied to factory production (Hopp and Spearman, 2001), built environment processes are arguably more complex. Nonetheless, work is under way towards identifying critical flows that may ultimately be amenable to mathematical analysis and modelling (Bertelsen *et al*, 2006, 2007).

It is worth noting that modelling applications have much in common with metaphors. If we accept that, whatever their basis, models represent simplifications of a system, then models are not so different from metaphors, which at their core are meant to be symbolic representations of something else. Indeed models (and diagrams) are aids to solving problems, not solutions in themselves. Thus, another way of viewing the work of Mittleton-Kelly (2003) or Gershenson and Heylighen (2005) is as outlining the concepts necessary to the creation of such models.

SIMPLE COMPLEXITY AND COMPLICATING SIMPLICITY

However, in some cases, the appearance of complexity may be generated by the inadequacy of our (non-mathematical) analytic methods. This can be one consequence of treating human organizational phenomena as amenable to exhaustive causal analysis. The use of mental terms (for example, disposition terms such as mood, attitude, personality, culture) in a non-causal analysis can provide simpler formulations. Moreover, metaphors can be misleading and it is worth looking more closely at how they are used in practice, i.e. in what ways they help, how they break down and with what consequences. For instance, Gershenson and Heylighen (2005) use the metaphor of complexity to attempt an escape from the restrictive bonds of Cartesian thought:

> *From the cybernetic perspective, there is no strict boundary between the 'material' and the 'mental' components: the information stored in my notebook is as much a part of my memory as is the information stored in my brain, and my email connection is as much part of my communication equipment as the nerves that control my speech. (Gershenson and Heylighen, 2005: 56)*

However, the new conceptual space created is relatively minor, compared with that achieved by phenomenological and linguistic philosophers (Wittgenstein, 1958; Schutz, 1962; Ryle, 1963). Thus, they find it necessary to resort to the Cartesian dualism of a universal objective reality and local subjective perception, in order to explain differences in perception. This, in turn, allows them to cling to the familiarity of a clockwork Newtonian universe, comforting though concealed, but it leaves the residual problem of how objective (physical) reality is to be reconciled with subjective (mental) perception (Button *et al*, 1995). There is a vast literature on this problem, which has proved an intractable theoretical inconvenience. It is most easily removed by recognizing that a) objectivity is itself a local achievement by situated perceptive beings and that b) the relaxation of distinctions is a linguistic move, rather than an ontological discovery.

The concept of dissipative structures may also be misleading. Although history is an important consideration in human systems, it would seem that complexity thinking has little to offer to an already highly developed field. In directing attention to the relatively impoverished concept of history employed in chemistry, Mittleton-Kelly (2003) runs the risk of overlooking the rich analysis that already exists in the sociological (Weber, 1976; Parsons, 1951) and management traditions (e.g. Kanter, 1985).

Wolfram's (2002) discovery that computer programs operating simple rules can generate highly complex patterns provides for an ambiguity in the description of such programs. Viewed as sets of rules, the programs may be described as simple; viewed as generators of outcomes, they may be described as complex. (Note also that the rules look simple because of the way they are stated, using boxes (Wolfram, 2002: 27) they would no doubt look more complicated if stated exclusively in English.) It is apparent that the perception of complexity depends on how a given mathematical phenomenon is viewed. Using the metaphor of the disturbances produced by two speedboats whose paths continuously interweave on an otherwise placid sea, Burke (1969) explicated such a phenomenon with regard to human behaviour; the complex or chaotic outcomes can be seen as the interaction of only two elements. The fact that complexity can be simple is confirmed by Gershenson and Heylighen (2005) who state that complexity can consist of as little as two parts. Sacks (1972) made a similar observation, speculating that it may not be that human behaviour is complicated that makes it so baffling to social scientists, but the fact that human behaviour is self-characterizing.

Sacks (1963) had previously employed the elaborate metaphor of a 'commentary machine' to illustrate the problem of social research. This machine functions to produce a commentary on its own activities. These are observable and in principle describable by social scientists, but these descriptions have the same ontological status as the descriptions produced by the machine itself. Indeed, they are often descriptions of descriptions. This presents problems related to both, i.e. how naturally produced descriptions are to be scientifically treated, and the nature of the claims that can be made for social scientists' own descriptions. The working out of these conundrums has been one source of the principle of unique adequacy (UA) discussed below. Note that the two parts here are the scientist (observer) and the subject (observed) – the complexity is in the relationship between them. It would be mistaken to believe that the two parts are, for instance, the action and its self-characterization, as Ryle's (1963) critique of Cartesianism clearly shows.

Snowden's Cynefin framework (Snowden, 2002; Kurtz and Snowden, 2003) is, among other things, an attempt to avoid such inappropriate categorizations of the world. Rather than categorizing the world directly, it offers a categorization of ways of looking at the world, with the additional proviso that the category boundaries are permeable and flexible. Thus, Cynefin provides a tool for achieving negotiated order (Strauss *et al*, 1964). Such tools provide a means of both understanding and creating systems in which:

> *the possibility that interactions between large numbers of entities, each entity responding to others on the basis of its own local organizing principles, will produce coherent patterns with the potential for novelty in certain conditions. (Stacey, 2001: 93)*

It may be that reaching such an understanding depends upon conducting research to meet the UA requirement of methods (Garfinkel and Wieder, 1992; Garfinkel, 2002). Livingston (1987) illustrates the significance of this requirement with reference to the patterns of movement formed by crowds crossing the road at a busy intersection. Overhead videotape footage reveals that crowds typically make wedge-shaped formations. There would seem to be three alternative strategies for answering the question as to how these formations are generated:

- use the videotape material to take an overview of the whole system and analyse the flows as they occur to determine possible chains of causality
- analyse the behaviour of single entities in order to discover their local organizing principles. This second course is clearly appropriate to complex systems
- however, for human systems, a third possibility exists. We may interview participants in crowds, or participate in crowds ourselves, in order to obtain an inside view of how the patterns are generated. In this way, the *decisions* of the crowd's members provide an additional rich source of whose complexity the UA requirement is designed to address.

It is noticeable that complexity-oriented scholars in the built environment have, to varying degrees, adopted this third strategy, the analysis of perceptions from particular identified points of view (Smallwood, 2000; Bertelsen, 2004; Bertelsen and Emmitt, 2005; Bresnen *et al*, 2005; Elhag *et al*, 2005).

The UA requirement has two forms, both of which:

> *are founded on the principle that the activities and procedures of persons in a setting can best be accounted for in terms of the understandings that those persons have of that setting. Thus, the task of the ethnographer is to render a clear description of those understandings, rather than offer an explanatory theory. (Rooke et al, 2004: 656)*

In its weak form, the UA requirement demands that 'the analyst must be *vulgarly* competent in the local production and reflexively natural accountability of the phenomenon' (Garfinkel and Wieder, 1992: 182). Thus, to analyse a decision-making process adequately, we must know what any participant in the process would ordinarily know about that process. This knowledge, expressed as competence, is the kind referred to by Ryle (1963) as 'knowing how'. In this form, the requirement is proposed as a criterion for adequate ethnography, the most certain method for acquiring such knowledge being participant observation. However, it is possible to usefully apply it to other forms of enquiry, such as interviews and questionnaires. Thus, for instance, a questionnaire designed by someone who had no direct knowledge of the process under study is likely to contain irrelevant, misleading or even meaningless questions.

The strong requirement concerns the reporting of research. It demands that the methods of analysis used to describe a process should be derived directly from that process. This assumes that the methods that participants use in the process of making a decision are sufficient to the purpose of producing an account of that decision. It is proposed as a standard for the description of any phenomenon which is composed of (and by) conscious beings who are able to produce an account of their own activities. It is proposed that methods of analysis that are alien to the analysed setting (thus introducing a theoretical spin to the description) must involve some distortion. This requirement is proposed as a criterion for ethnomethodology. However, reports have been written outside this discipline which, at least in part, meet this form of the requirement (e.g. Dutton and Starbuck, 1971). It currently informs much of the work being done in the field of Computer Supported Co-operative Work (CSCW). For instance, Button and Sharrock (2002) have offered a UA description of the routine decision-making processes of a production scheduler in a print works.

Intrinsic complexity could be said to lay in several areas of the activity, for instance, the craft-like nature of the printing process; variability in the capability and performance of printing machines; conflicts between the technology used by the design house customers and the ability of the printing equipment to replicate this. Intrinsic complexity could also be seen to be inherent in the relationships between the customer and the printers, since these are taken as a given, although actual instantiations of uncertainties and 'buggeration' factors may vary from case to case.

Induced complexity would seem to emanate from the nature and variability of the work-breakdown structures within the printing organization; the relationships between the various roles of e.g. account holder, scheduler, shop floor personnel, management; the range and depth of both the formal and tacit knowledge and experience of the individual scheduler; the variation and evolution of organizational units from job to job and shift to shift.

Decision-making in this context is affected by both the above and by certain elements of what could be called 'emergent behaviour' or 'system interoperability'. The former, because any combination of type of job, different team, different shift, different machines, degree of foreseeable problems, etc. will result in a potential event (e.g. job stopped, job delayed, etc.) which will require decisions to be made on the schedule to be regularly revisited in a partially structured *ad hoc* manner. The latter, because the overall printing capability is not an integrated or interoperable system and perhaps, given its nature, could never be. In a sense, the scheduler functions to integrate the system, often acting on the fly. It is not clear how much of the decision-making process is repeated in terms of process, variables taken into account, and people consulted. By this we mean: does a set of decision-making processes and variables come into play for similar sets of problems, or is it a totally random process that varies from scheduler to scheduler?

Although no similar studies exist as yet of the design, construction or maintenance of the built environment, there are many parallels that may be drawn. For instance, the role of the scheduler as an absorber of inter-task complexity is highly relevant to the sometimes seemingly intractable problems of co-ordination on construction projects. Some key decision-making techniques are:

- to eliminate the need to consider alternatives (although possibly holding these in reserve) by producing a single workable schedule
- to simplify the interactions between tasks to a smaller set of known ones (while possibly retaining others mentally)
- to leave intra-task complexities, that will take effect once the task commences, and may, via the inter-task interactions, affect other tasks, to be fixed by the task operators, based on their knowledge and experience.

CONCLUSION

We have identified two forms of complexity thinking – metaphor and modelling. The immediately apparent use of metaphor in the examples we have considered is rhetorical, as suits the use of a poetic device. Complexity metaphors can be used to call attention to important features of organization and to advocate particular styles of management.

We have observed that the modelling form is sometimes mistakenly thought to enable prediction. It should rather be thought of as a means of exploring possibilities, particularly those that would otherwise not be apparent.

We have also noted what models and metaphors have in common – both are forms of analogy. Whereas metaphor offers a direct comparison between phenomena found in the world (such as an organization and a machine), a model constitutes a deliberately simplified representation of a phenomenon. Thus, models are a way of relating mathematical techniques to actual existing systems – we have identified several applications both in the built environment and more widely.

The two approaches offer very different resources for tackling a problem. The model, through simplification, may help us to identify the important or essential features of a system, excluding obfuscating detail. By contrast, a metaphor adds something new, enriching our understanding to provide a further possibility of creative solution. The model may be said to sharpen our focus, the metaphor to broaden it. Together, they can form a bridge for the introduction of complex mathematical techniques. Metaphors may provide a common ground for explanation of the maths – a common practice in mathematical education. Models are necessary to reduce a phenomenon to a mathematically analysable form.

However, we must not forget that a solution to a maths problem is simply that. Its applicability to the real world depends on the constitution of the models we have built. And complexity theory tells us that patterns can emerge where we least expect them. We must be sure we are clear about the questions we are asking and must be cautious in our assumptions that we have identified the important variables. We should bear in mind that, as yet, the most successful applications of flow theory in the management of design and construction processes – the Last Planner and ADePT tools – have a limited mathematical content.

Furthermore, it is important to accept and understand that different people have different mental models and, as such, will view metaphors in different ways. We have noted some ways in which applications of complexity theory to organizational analysis may be misleading. It is important in avoiding these pitfalls that we consider the ways in which complexity is being used. The notion of history, for instance, may be important in identifying a quality of systems that it may then be possible for mathematicians to model. It is hardly likely in itself to add anything to social science understanding.

Finally, we have considered a metaphor arising not from complexity theory, but from social science. The metaphor of the commentary machine leads us away from rhetoric, towards specification of the actual psycho-social processes that constitute the stuff out of which real-life organizations are built. We have shown how reports of such specifications can reveal actual features of complex organizational settings.

ACKNOWLEDGEMENTS

This paper was written under the aegis of the Knowledge and Information Management (KIM) Through-Life Grand Challenge Project (www.kimproject.org) funded primarily by the Engineering and Physical Sciences Research Council (EPSRC – Grant No. EP/C534220/1), the Economic and Social Research Council (ESRC – Grant No. RES-331-27-0006), Loughborough University's Innovative Design and Manufacturing Research Centre (IMRC – Grant No. EP/E002323/1) and SCRI, Salford University's Innovative Design and Manufacturing Research Centre (IMRC – Grant No. EP/E001882/1).

AUTHOR CONTACT DETAILS

John Rooke: SCRI, Institute for Research in the Built and Human Environment, School of the Built Environment, University of Salford, 4th Floor, Maxwell House, The Crescent, Salford M5 4WT, UK. Tel: +44 (0)161 295 6344, fax: +44 (0)161 295 4587, e-mail address for correspondence: j.rooke@salford.ac.uk
Ella-Mae Molloy: e-mail: E.Molloy@lboro.ac.uk
Murray Sinclair: e-mail: M.A.Sinclair@lboro.ac.uk
Lauri Koskela: e-mail: L.J.Koskela@salford.ac.uk
Mohan Siriwardena: e-mail: mohansiri@yahoo.com
Mike Kagioglou: e-mail: m.kagioglou@salford.ac.uk
Carys Siemieniuch: e-mail: C.E.Siemieniuch@lboro.ac.uk

REFERENCES

Allen, P.M., 1997a, *Cities and Regions as Self Organising Systems*, Vol 1, Amsterdam, Gordon and Breach.

Allen, P.M., 1997b, 'Cities and regions as evolutionary, complex systems', in *Geographical Systems*, 4(1), 103–130.

Arthur, B.W., 1990, 'Positive feedback in the economy', in *Scientific American*, 80(1), 92–99.

Arthur, B.W., 1995, *Increasing Returns and Path Dependence in the Economy*, Ann Arbour, University of Michigan Press.

Austin, S., Baldwin, A., Baizhan, L.I. and Waskett, P., 2000, 'Analytic design planning technique (AdePT): a dependency structure matrix tool to schedule the building design process', in *Construction Management and Economics*, 18(2), 173–182.

Austin, S., Baldwin, A. and Newton, A., 1994, 'Manipulating the flow of design information to improve the programming of building design', in *Construction Management and Economics*, 12(5), 445–455.

Baldwin, A.N., Shen, L.Y., Poon, C.S., Austin, S.A. and Wong, I., 2008, 'Modelling design information to evaluate pre-fabricated and pre-cast design solutions for reducing construction waste in high rise residential buildings', in *Automation in Construction*, 17(3), 333–341.

Ballard, G., 2000, *The Last Planner System of Production Control*, PhD thesis, School of Civil Engineering, Faculty of Engineering, The University of Birmingham, UK, http://www.leanconstruction.org/pdf/ballard2000-dissertation.pdf

Ballard, G. and Howell, G., 1997a, 'Implementing lean construction: stabilizing work flow', in L. Alarcon (ed), *Lean Construction*, Rotterdam, A.A. Balkema, 101–110. Paper originally presented (1994) at the *International Seminar on Lean Construction,* Edefica, Pontificia, Universidad Catolica, Santiago de Chile.

Ballard, G. and Howell, G., 1997b, 'Implementing lean construction: improving downstream performance', in L. Alarcon (ed), *Lean Construction*, Rotterdam, A.A. Balkema, 111–126.

Ballard, G. and Howell, G., 1998, 'Shielding production: essential step in production control', in *Journal of Construction Engineering and Management*, 124(1), 11–17.

Berry, J.S., 1996, *Introduction to Non-Linear Systems*, London, Arnold.

Bertelsen, S., 2004, 'Construction management in a complexity perspective', in *1st International SCRI Symposium, Salford Centre for Research and Innovation (SCRI) in the Built and Human Environment, University of Salford, March 2004*, 346–355, http://www.bertelsen.org/strategisk_r%E5dgivning_aps/pdf/Salford%202004.pdf

Bertelsen, S. and Emmitt, S., 2005, 'The client as a complex system', in I. First (ed), *Proceedings of the 13th Annual Conference of the International Group for Lean Construction, International Group for Lean Construction, Sydney, Australia,* 73–80, http://www.iglc.net/conferences/2005/papers/

Bertelsen, S., Henrich, G., Koskela, L. and Rooke, J., 2007, 'Construction physics', in C.L. Pasquire and P. Tzortzopoulos (eds), *Proceedings of the 15th Annual Conference of the International Group for Lean Construction, Michigan State University, East Lansing,* 13–26, http://www.iglc.net/conferences/2007/folder.2007-06-29.2095743756/

Bertelsen, S., Koskela, L., Henrich, G. and Rooke, J., 2006, 'Critical flow; towards a construction flow theory', in C.L. Pasquire and P. Tzortopoulos (eds), *Proceedings of the 14th Annual Conference of the International Group for Lean Construction, Pontificia Universidad CatÛlica de Chile, Santiago de Chile, July 2006*, 31–40, http://www.iglc2006.cl/papers.html

Boussabaine, A.H. and Elhag, T., 1999, 'Applying fuzzy techniques to cash flow analysis', in *Construction Management and Economics,* 17(6), 745–755.

Boussabaine, A.H., 1996, 'The use of artificial neural networks in construction management: a review', in *Construction Management and Economics*, 14(5), 427–436.

Bresnen, M., Goussevskaia, A. and Swan, J., 2005, 'Organizational routines, situated learning and processes of change in project-based organizations,' in *Project Management Journal,* 36(3), 27–41.

Burke, K., 1969, *A Grammar of Motives*, London, University of California Press.

Button, G. and Sharrock, W., 2002, 'Operating the production calculus: ordering a production system in the print industry', in *British Journal of Sociology*, 53(2), 275–289.

Button, G., Coulter, J., Lee, J.R.E. and Sharrock, W.W., 1995, *Computers Minds and Conduct*, Cambridge, Polity Press.

Carroll, T. and Burton, R.M., 2000, 'Organisations and complexity: searching for the edge of chaos', in *Computational and Mathematical Organisation Theory*, 6(4), 319–337.

Chen, J., Taong, L. and Ngai, E.W.T., 2007, 'Inter-organizational knowledge management in complex product systems: challenges and an exploratory framework', in *Journal of Technology Management in China*, 2(2),134–144.

Codd, E.F., 1968, *Cellular Automata*, New York and London, Academic Press.

de Bono, E., 1971, *The Use of Lateral Thinking: Break the Stranglehold of Logical Thinking*, Harmondsworth, Penguin.

Dijkstra, J., Jessurun, J. and Timmermans, H., 2001, 'A multi-agent cellular automata model of pedestrian movement', in M. Schreckenberg and S.D. Sharma (eds), *Pedestrian and Evacuation Dynamics*, Berlin, Springer-Verlag, 173–181.

Dutton, J.M. and Starbuck, W., 1971, 'Finding Charlie's run-time estimator', in J.M.S. Dutton and W. Starbuck (eds), *Computer Simulation of Human Behaviour*, Wiley, Chichester, 218–242.

Elhag, T.M.S., Boussabaine, A.H. and Ballal, T.M.A., 2005, 'Critical determinates of construction tendering costs: quantity surveyors' standpoint,' in *International Journal of Project Management,* 23(7), 538–545.

Gardner, M., 1970, 'The fantastic combinations of John Conway's new solitaire game "life",' in *Scientific American*, 233, 120–123.

Garfinkel, H., 2002, *Ethnomethodology's Program: Working out Durkheim's Aphorism*, Lanham, Rowman and Littleford.

Garfinkel, H. and Wieder, D.L., 1992, 'Two incommensurable, asymmetrically alternate technologies of social analysis', in G. Watson and R.M. Seiler (eds), *Text in Context*, London, Sage, 175–206.

Gershenson, C.H. and Heylighen, F., 2005, 'How can we think complex?', in K.A. Richardson (ed), *Managing Organizational Complexity: Philosophy, Theory and Application*, Information Age Publishing, 47–61.

Groak, S., 1994, 'Is construction an industry? Notes towards a greater analytic emphasis on external linkages', in *Construction Management and Economics,* 12(2), 287–293.

Guastello, S.J., 1995, *Chaos, Catastrophe and Human Affairs: Applications of Nonlinear Dynamics to Work, Organisations and Social Evolution*, Mahwah, New Jersey, Lawrence Erlbaum Associates.

Handy, C., 1994, *The Empty Raincoat: Making Sense of the Future*, London, Hutchinson.

Hofstede, G., 2002, *Culture's Consequences: Comparing Values, Behaviours, Institutions and Organizations Across Nations*, London, Sage.

Hopp, W.J. and Spearman, M.L., 2001, *Factory Physics*, New York, McGraw-Hill.

Ivory, C. and Alderman, N., 2005, 'Can project management learn anything from studies of failure in complex systems?' in *Project Management Journal,* 36(3), 5–16.

Kanter, R.M., 1985, *The Change Masters: Corporate Entrepreneurs at Work*, London, Unwin.

Kirkham, R.J. and Boussabaine, A.H., 2005, 'Forcasting the residual service life of NHS hospital buildings: a stochastic approach', in *Construction Management and Economics*, 23(5), 521–529.

Koskela, L., 1992, *Application of the New Production Philosophy to Construction*, Technical Report No 72, Center for Integrated Facility Engineering, Stanford University, California.

Koskela, L., 2000, *An Exploration towards a Production Theory and its Application to Construction,* VTT Publication 408, VTT Technical Research Centre of Finland.

Kosko, B., 1993, *Fuzzy Thinking – The New Science of Fuzzy Logic,* 1st edn, London, Flamingo.

Kurtz, C.F. and Snowden, D.J., 2003, 'The new dynamics of strategy: sense-making in a complex and complicated world', in *IBM Systems Journal*, 42(3), 462–483.

Livingston, E., 1987, *Making Sense of Ethnomethodology*, London, Routledge and Kegan Paul.

McCarthy, I., Tsinopoulous, C., Allen, P. and Rose-Anderssen, C., 2006, 'New product development as a complex adaptive system of decisions', in *Journal of Product Innovation Management*, 23(5), 437–456.

Mittleton-Kelly, E., 2003, *Complex Systems and Evolutionary Perspectives on Organisations: The Application of Complexity Theory to Organisations*, Oxford, Elsevier.

Morgan, G., 1986, *Images of Organization*, London, Sage.

Parsons, T., 1951, *The Social System*, London, Routledge and Kegan Paul.

Parsons, T., 1968, *The Structure of Social Action*, London, Collier Macmillan.

Payr, S. and Trappl, R., 2004, *Agent Culture: Human-Agent Interaction in a Multicultural World*. London, Lawrence Erlbaum Associates.

Peters, T.J. and Waterman, R.H., 1982, *In Search of Excellence*, London, Harper and Row.

Richardson, K., Mathieson, G. and Cilliers, P., 2000, 'The theory and practice of complexity science – epistemological considerations for military operational analysis', in *SysteMexico*, 1, 25–66.

Robertson, D.A., 2005, 'Agent-based models to manage the complex', in K.A. Richardson (ed), *Managing Organisational Complexity: Philosophy, Theory and Application*, Charlotte, NC, Information Age Publishing, 419–432.

Rooke, J., Seymour, D. and Fellows, R., 2004, 'Planning for claims; an ethnography of industry culture', in *Construction Management and Economics*, 22(6), 655–662.

Ryle, G., 1963, *The Concept of Mind*, Harmondsworth, Penguin.

Sacks, H., 1963, 'Sociological description,' in *Berkely Journal of Sociology*, 8, 1–16.

Sacks, H., 1972, 'An initial investigation of the usability of conversational data for doing sociology', in D. Sudnow (ed), *Studies in Social Interaction*, New York, Free Press, 31–74.

Schutz, A., 1962, *The Problem of Social Reality*, The Hague, Marinus Nijhoff.

Seese, D. and Schlottman, F., 2003, 'The building blocks of complexity: a unified criterion and selected applications in risk management', in *Modelling, Computing and Mastering Complexity Series,* Paper 14, Society for Computational Economics.

Shen, W., Norrie, D. and Barthers, J., 2001, *Multi-agent Systems for Concurrent Intelligent Design and Manufacturing*, New York, Taylor and Francis.

Smallwood, J., 2000, 'Contractor performance: clients' perceptions', in A.B. Ngowi and J. Ssegawa (eds), *Challenges Facing the Construction Industry in Developing Countries, Proceedings of the 2nd International Conference of the CIB Task Group 29 (TG29) on Construction in Developing Countries, CIB, Gaborone, Botswana, November 2000*, 128–138.

Snowden, D., 2002, 'Complex acts of knowing: paradox and descriptive self-awareness', in *Journal of Knowledge Management*, 6(2), 100–111.

Stacey, R.D., 2001, *Complex Responsive Processes in Organizations: Learning and Knowledge Creation*, London, Routledge.

Steeb, W.-H., 2005, *The Nonlinear Workbook,* 3rd edn, London, World Scientific.

Strauss, A., Schatzman, L., Bucher, R., Ehrlich, D. and Sabshin, M., 1964, *Psychiatric Ideologies and Institutions*, New York, The Free Press of Glencoe.

Thomassen, M.A., Sander, D., Barnes, K.A. and Nielsen, A., 2003, 'Experience and results from implementing lean construction in a large Danish contracting firm', in J.C. Martinez and C. Formoso (eds), *Proceedings of the 11th Annual Conference on Lean Construction, Virginia Tech, Blacksburg, Virginia, July 2003*, 73–85, http://strobos.cee.vt.edu/IGLC11/PDF%20Files/54.pdf

Vakili-Ardebili, A. and Boussabaine, A.H., 2007, 'Application of fuzzy techniques to develop an assessment framework for building design eco-drivers', in *Building and Environment,* 42(11), 3785–3800.

von Neumann, J., 1966, *Theory of Self-Reproducing Automata*, London, University of Illinois Press.

Weber, M., 1976, *The Agrarian Sociology of Ancient Civilizations*, London, NLB.

Wittgenstein, L., 1958, *Philosophical Investigations*, Oxford, Blackwell.

Wolfram, S., 2002, *A New Kind of Science*, Illinois, Wolfram Media.

ARTICLE

Feng shui: an alternative framework for complexity in design

Michael Y. Mak and S. Thomas Ng

Abstract

Building design is a very complex process that requires an integration of creativity, rationality and self-organization. Although there are numerous studies that examine how architects learn and apply their knowledge to design new buildings, the design process still cannot be explicitly defined. Because the way that architectural design knowledge is applied to building design cannot be formulated, the term 'black box' is used to represent an architect's design mechanism that cannot be logically explained. Designers use intuition and individual past experiences to make heuristic decisions when tackling new situations in building design. These characteristics are similar to the Chinese philosophy of direct insight into the nature and principles and practice of feng shui as applied to building design. Feng shui is traditional Chinese wisdom aimed at creating harmony among the environment, buildings and people. The aims of this paper are to examine the potential of feng shui knowledge as an alternative approach to building design and to derive a conceptual framework based on the principles and practice of feng shui.

■ *Keywords* – Feng shui; building design; complexity in design

INTRODUCTION

The nature of design falls between science and art, and architectural design is a complex and intuitive process. Architects use knowledge from past experiences and from their own theories to create spaces with desired experiential qualities (Suckle, 1980). Jones (1970) recognized the complexity of the design process and suggested that a designer working on a problem is engaged simultaneously in three types of activity:

- Creativity: a designer is a 'black box' out of which comes a mysterious creative leap.
- Rationality: a designer is a 'glass box' inside which can be discerned a completely explicable rational process.
- Control over the design process: a designer is a self-organizing system capable of finding short-cuts across unknown territory.

This integrative design process is similar to the Chinese holistic view and the feng shui approach to the built environment.

Feng shui is a body of ancient Chinese wisdom in knowledge and experience relating to the built environment that has accumulated over more than 3000 years. There are two main schools of thought and practice in feng shui – the Compass School and the Form School. The Form School approach has been well recognized and widely accepted by feng shui researchers as comprising a scientific basis for the analysis of the built environment (He, 1990; Cheng and Kong, 1993). The principles and practices of feng shui are aimed at creating a harmonized built environment for people to live in, and it represents a traditional Chinese architectural theory for selecting favourable sites as well as a theory for designing cities and buildings (Lee, 1986).

Nowadays, as many researchers seek to establish a deeper understanding of the relationships between

doi:10.3763/aedm.2008.S307 ISSN: 1745-2007 (print), 1752-7589 (online) www.earthscanjournals.com

the human and natural environments, architects begin to recognize feng shui as a broad ecologically and architecturally connected paradigm. Hwangbo (1999) believed that the practice of feng shui is an intuitive matter involving site selection and spatial organization, and it has strong parallels with the Western concept of geometry in architectural design.

The aims of this paper are to examine the potential of feng shui knowledge as an alternative approach to building design and to derive a conceptual framework based on the principles and practice of feng shui which comprises the following steps. First, the nature and complexity of design are explored, and similarities to Chinese philosophy and feng shui knowledge are explained. Second, the principles and practices of the Form School approach are introduced because it has been well recognized and widely accepted by feng shui researchers as comprising a scientific basis for the analysis of the built environment. Third, four fundamental concepts of the Form School approach are derived which provide a broad understanding of feng shui knowledge. Finally, a conceptual framework of feng shui for building design is constructed utilizing the concept hierarchy approach to establish a feng shui knowledge structure.

The findings of this paper contribute significantly to current practice of architectural design by providing an alternative framework for building design and creating a hierarchical structure of feng shui knowledge for building design. The purpose of this paper is not to give feng shui knowledge validity from a Western perspective of science and does not set out to challenge conventional theories of architectural design; rather, this feng shui design framework suggests the significance of an unconventional alternative system of analysis of building design.

COMPLEXITIES IN DESIGN

It is generally recognized that design falls somewhere between two extremes – science and art – and architectural design lies near the centre of this spectrum (Jung, 1996). Sargent (1994) studied the nature of design and suggested that there can be no unitary 'science of design'.

In the 1960s, design was interpreted in a similar way to the interpretation of science – the format of a scientific method was constructed in the form of analysis and synthesis. This approach originated in the notions of objectivity, rationality and universalism that are believed to constitute the scientific method – breaking the problem down to fragments and solving each of these separately before attempting some grand synthesis (Asimow, 1962; Jones, 1963; Archer, 1969).

However, more researchers (Gregory, 1966; Simon, 1969) began to recognize that design is not like science. This approach was largely based on the philosophy of science developed by Popper (1972) and Kuhn (1970). Popper (1972) suggested the concept of the 'conjecture–refutation' model of the scientific method instead of the logical process of induction whereby laws can be inferred from true facts. Kuhn (1970) proposed a 'paradigm' concept to explain the evolution of scientific knowledge, where scientific knowledge progresses through a 'paradigm shift'.

Jones (1970) admitted that it is impossible to ascertain the nature of design. He outlined three key points to attempt to describe the nature of design, which are summarized by Raftery (1991) as follows:

- There are very often long periods when the person who is about to make an original work seems to do nothing except take in information and labour fruitlessly at seemingly trivial aspects of a problem. This is known as 'incubation'.
- The solution to a particular problem or the occurrence of an original idea often happens when, at some particular point in time, everything seems to fall into place. This is known as the 'leap of insight' or 'change of set'. Basically the problem is perceived in a new light and very often an apparently complex problem turns into a very much simpler one.
- The main enemies of originality are mental rigidity and wishful thinking.

Incubation was a concept proposed by Poincare (1929) who interpreted the conditions for creativity as:

- A period of conscious work, data assembled, problem defined, etc. and some trials made at solutions.

- The unconscious mind works at useful and fertile combinations during this time and useless areas are inhibited.
- A hypothesis is derived which gives a fruitful direction. In Poincare's own words as 'a period of preliminary conscious work always precedes all fruitful unconscious work'.

Following the conclusions of Jones' and Poincare's work, design was described as a 'problem-solving' activity. However, the solution to the problem was perceived by individual designers and not measured objectively. The further implications of both Jones' and Poincare's work were that the solution is produced by an unconscious thought process. Humphreys (1976) recognized that this mystical approach is remarkably similar to the concept of 'satori' (the enlightenment in Zen Buddhist thought) in Chinese philosophy.

FORM SCHOOL APPROACH

Feng shui is a body of Chinese wisdom in knowledge and experience relating to the built environment that has accumulated over more than 3000 years. It has been developing throughout the existence of Chinese history and civilization, and has evolved from Chinese philosophy. It is founded on the earliest and greatest document of Chinese philosophy, *Yi Jing* (the *Book of Changes*) that originated around 800 BC. The primitive knowledge of feng shui was based on observations from three sources – astronomical phenomena, natural phenomena and human behaviour (Feuchtwang, 1974). The principles and practices of feng shui are aimed at creating a harmonized built environment. Feng shui represents a traditional Chinese architectural theory for selecting favourable sites as well as a theory for designing cities and buildings (Lee, 1986).

There are two main schools of thought and practice in feng shui – the Compass School and the Form School. The Compass School is based on metaphysical speculations of cosmology, in particular by analysing the directional aspects in terms of the relationships between the five elements, eight trigrams, heavenly stems, earthly branches and constellations. Practice in the Compass School uses primarily the Luopan (feng shui compass) and the composed elements of time in space (Skinner, 1982; Chiou and Krishnamurti, 1997). The Form School is primarily based on the verification of the physical configuration of mountains and watercourses surrounding sites and buildings. Its theory was built upon an understanding of the landscape – the profiles of the land, the sources of rivers and the terrain. The practice of the Form School first observes the land formation and terrain, and then determines the location and orientation of buildings.

The development of the Form School was widely accepted by the upper class of the ancient Chinese society and attracted scholars and intellectuals to join its practice. The principles of the Form School have been applied to design and construct castles, palaces and towns in China since ancient times (He and Luo, 1995). Lee (1986: 367) suggested that the principles and practices of the Form School approach represent 'a compendium of Chinese architectural theory'.

Since the Ming Dynasty (1368–1644), these two schools of thought were not exclusively attached to their own methods for the practice of feng shui, but rather combined and integrated ideas from both (Lee, 1986). However, the Form School approach remained the primary consideration in feng shui practice (Xu, 1990; Too, 1996). Contemporarily, the Form School approach has been recognized as comprising a scientific basis for the analysis of the built environment (He, 1990; Wang, 1992; Cheng and Kong, 1993; Mak and Ng, 2005). For instance, research investigations carried out by Xu (1990) have compared the feng shui concepts using the Form School approach and the Hendler model, a well-known Western model of site analysis. The results indicated that feng shui is a more powerful tool in site analysis than the Hendler model.

According to *Zang Shu* (the *Book of Burial*), the first surviving important literature on Form School was written by Guo Pu (276–324). There are five main theories in terms of Form – Qi, Wind-water, Four Emblems, Form and Direction (He, 1990). The Form School approach considers mountain ridges, surrounding hills, watercourses, locations and orientations as the most important terrestrial and celestial elements for human dwellings because these elements represent both terrestrial and celestial Qi. These elements comprise the basic terms of the Form School approach and are known as the 'five feng shui geographical secrets', i.e. dragon, sand, water, cave and direction (Lip, 1979):

- dragon – means the mountain ridges to be traced, and represents the topography
- sand – means the enfolding hills and soil condition, and represents the surrounding environment
- water – means the flow of water through or bypassing the site
- cave – or 'feng shui spot' means the niche position, and represents the best location
- direction – means the facing direction of the site and building, and represents the orientation.

RESEARCH METHODOLOGY

The development of a conceptual framework follows the procedure of the grounded theory approach. Glaser and Strauss (1967) described this approach as the discovery of theory from data, and emphasized its importance in providing researchers with predictive understanding, explanations, interpretations and applications. The procedures of developing the conceptual framework of feng shui knowledge are as follows:

- review and summarize relevant concepts from feng shui knowledge
- develop an appropriate framework for analysis
- review trends, patterns and issues in feng shui principles and theories
- reinterpret and develop knowledge hierarchy from the feng shui perspective in relation to building design
- integrate the conceptual framework development.

The fundamental concepts of the Form School approach are first derived from existing principles and practice. This process follows the procedure of grounded theory through the analysis of literature (Mak, 2004). Hence the hierarchies of feng shui concepts are identified and constructed in a form of hierarchical structure utilizing concept hierarchy approach. Concept hierarchy is used to organize factual domain knowledge and symbolic structural knowledge in the form of a generalization hierarchy. It is a common approach to utilize concept hierarchy for organizing structural knowledge and constructing the knowledge base because of its efficient mechanism to store and generalize a large body of interrelated concepts (Tam, 1993).

FOUR FUNDAMENTAL CONCEPTS OF THE FORM SCHOOL APPROACH

The five main theories of the Form School approach together with the five feng shui geographical secrets are developed into four fundamental concepts of the Form School approach to feng shui knowledge for building design:

- the concept of the feng shui model
- the concept of parallelism
- the concept of four design modules and
- the concept of feng shui design criteria.

CONCEPT OF THE FENG SHUI MODEL

The combination of these five feng shui geographical elements and the four emblems (green dragon, white tiger, black tortoise and red bird as the four cardinal directions) produce a classic feng shui model. This model has been interpreted in diagrams of spatial organization of auspicious mountains and watercourses in most of the ancient feng shui literature. Many feng shui researchers have summarized these diagrams into a simplified diagram of a feng shui model as shown in Figure 1 (Shang, 1992; Chung and Kong, 1993; Han, 1995; Yi *et al*, 1996; He, 1998). This diagram illustrates the relationships between the key elements of the five feng shui geographical secrets being considered and how dragon vein, four emblems in sand, water feature, cave and bright court, and their directions were integrated into a feng shui model.

CONCEPT OF PARALLELISM

Most feng shui scholars (Skinner, 1982; Lee, 1986; Xu, 1990) recognized that the theories and practices of feng shui work in both macrocosm and microcosm. In terms of feng shui, a building is considered an architectural as well as a cosmic structure (Lee, 1986). Territory on earth is organizationally analogous with the four quadrants in the celestial sphere. These four quadrants are called azure dragon, red bird, white tiger and black tortoise, and are commonly known as the 'four emblems'. Each of these heavenly

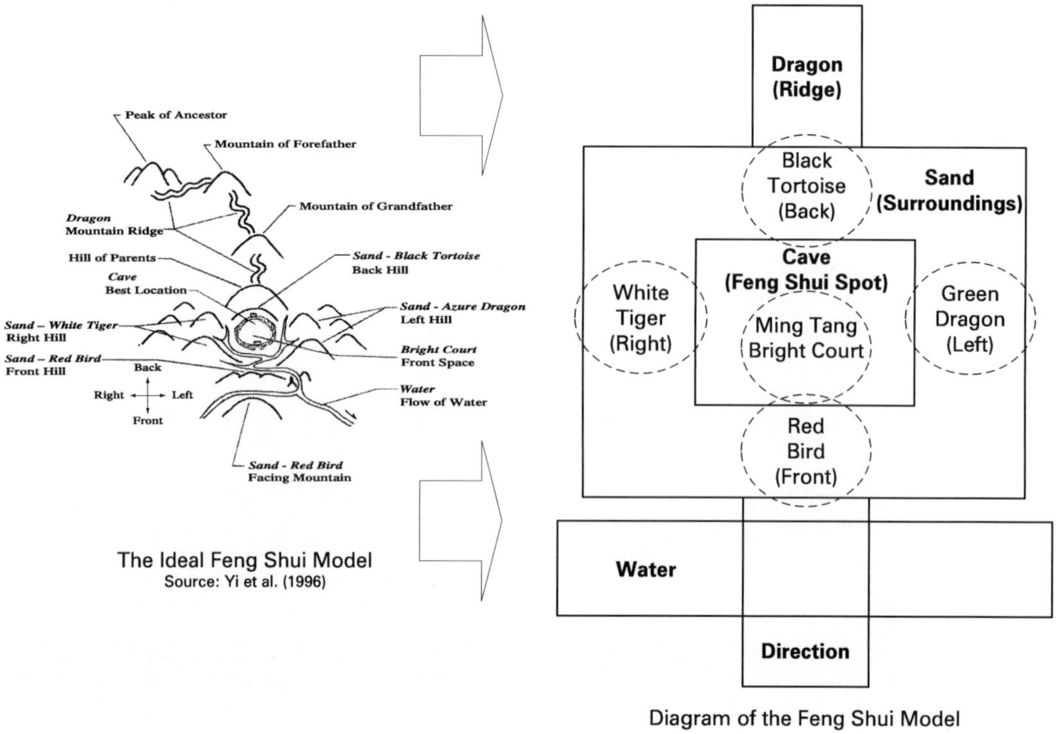

FIGURE 1 The feng shui model.
Source: Yi et al (1996) and Mak (2004)

quadrants is identified with the regions of east, south, west and north, respectively. This relationship is recognized as the concept of parallelism in feng shui (Lee, 1986). The concept of a feng shui model not only applies to landscape and site selection, but can also be applied to the interior layout of buildings. Therefore, whether it is dealing with physical or topographical elements, or a housing structure or the proportional relationships of the interior of a house, the same principles and relationships of the feng shui model are still applied as shown in Figure 2.

CONCEPT OF FOUR DESIGN MODULES

When describing the site conditions and the design of dwellings, most of the feng shui texts, such as *Yang Zhai Shi Shu* (*Ten Books on Dwellings of Living*), categorized these aspects into outer form and inner form. According to Lee (1986), the outer form can be identified as the location of the site, conditions that surround the site, topographical conditions of the site and the shape of the site. The inner form can be identified as the layout of the building, elevations of the building, and elements of the building. Feng shui scholars Cheng and Kong (1993) explained the application of the Form School approach to the design of dwellings and proposed a further classification into four design modules – surrounding environment, external layout, internal layout and interior arrangement (see Figure 3).

- Surrounding environment. This aspect looks into the surrounding environment from a geographical point of view. This includes natural elements of topography, geographical features, mountains,

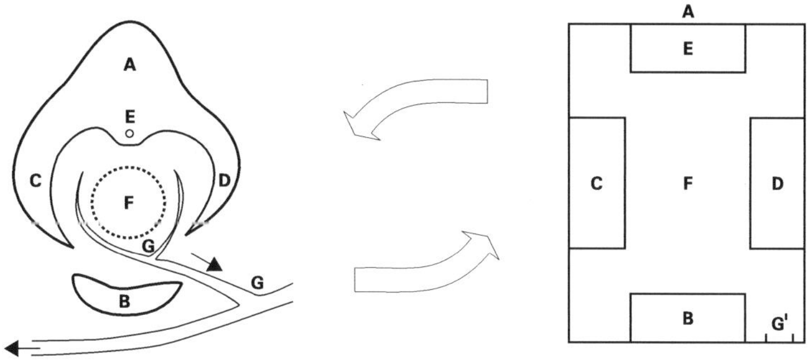

A Black Turtle
B Red Bird
C White Tiger
D Azure Dragon
E Feng Shui Spot
F Ming Tang
G Water Mouth
G' Gate

FIGURE 2 Correspondence of the four emblems in nature and architecture.
Source: adapted from Lee (1986)

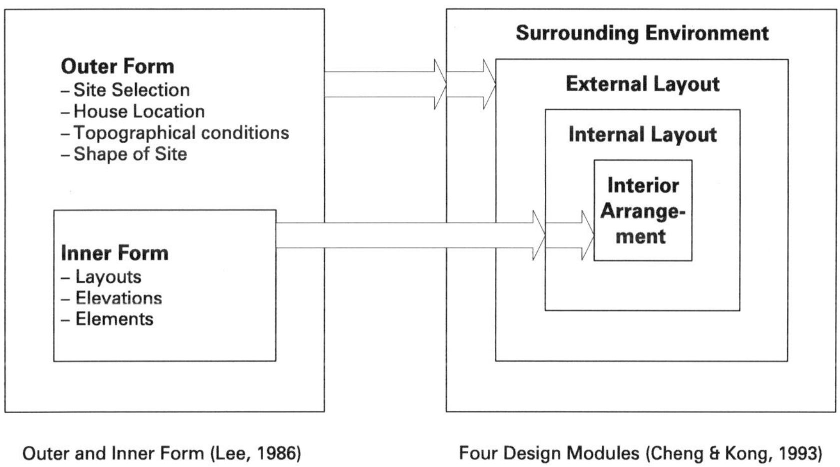

FIGURE 3 Four design modules.
Source: Mak (2004)

watercourses, directions, views and man-made structures of roads and adjacent buildings.

- External layout. The external layout deals with the external shape and exterior space of a building. This includes the shape of the site, geometry of the building, open space, entrances, driveways, landscaping and plants.
- Internal layout. The internal layout reflects the spatial management of a building. It considers the locations and functions of rooms, circulation patterns, and internal elements of structure, such as columns and beams, staircases, ceilings, doors and window openings (Rossbach,1987).

TABLE 1 24 key feng shui criteria for building design

SURROUNDING ENVIRONMENT	EXTERNAL LAYOUT	INTERNAL LAYOUT	INTERIOR ARRANGEMENT
• Topography	• Shape of site	• Layout	• Door openings
• Front of site	• Entrance	• Doors	• Bedroom
• Rear of site	• Shape of building	• Windows	• Kitchen
• Sides of site	• Orientation	• Shape of rooms	• Living room
• Street location	• Trees	• Staircase	• Bathroom
• Water view	• Pond	• Ceiling	
• Wind direction			

Source: Mak (2004)

- Interior arrangement. The interior arrangement addresses the internal room arrangement and furniture placement. It considers the size and proportion of rooms and windows and doors. Furniture placement is a major part of interior arrangement. In practice, the bed in the bedroom, the stove in the kitchen, the desk in the office etc. have substantial effects on the use of a building (Rossbach, 1987).

CONCEPT OF FENG SHUI DESIGN CRITERIA

Most contemporary feng shui scholars (Lip, 1979, 1986; Rossbach, 1984, 1987; Lee, 1986; Xu, 1990; Han, 1995; Choy, 1999) have set up their own criteria for feng shui design. For instance, Lee (1986) outlined three basic criteria for architectural design; Xu (1990) derived a four-step landscape model to deal with land formations; Han (1995) used 24 major criteria for selection of the best location; Lip (1979, 1986) listed a set of standard rules of thumb for assessment of architectural design; Choy (1999) suggested a 10-point design criteria checklist for property selection; and Rossbach (1984, 1987) provided a set of interior design diagrams for furniture placement. Although these criteria, derived from various contemporary feng shui scholars, were presented in different formats, they all follow the principles and practice of the Form School approach.

Based on these contemporary practices for feng shui design, 24 key criteria are identified (Table 1) and grouped according to the four design modules as shown in Figure 4.

FENG SHUI DESIGN FRAMEWORK

CONCEPT HIERARCHY APPROACH

Based on the four fundamental concepts of the Form School approach, i.e. the concept of the feng shui model, the concept of parallelism, the hierarchy of four design modules, and criteria for feng shui design, an overall view of the Form School approach was provided. It is necessary to integrate these four fundamental concepts into a single diagrammatical organization of the Form School approach as shown in Figure 5. This diagram indicates that these four fundamental concepts are integrated and interrelated, and no single concept should be considered in isolation. For instance, the concept of the feng shui model integrated with the concept of parallelism is reflected in the hierarchy of the four design modules, and the criteria for feng shui design corresponding to the concepts of the feng shui model is applied to the hierarchy of the four design modules.

Based on the organization of the Form School approach, it is necessary to formulate a suitable process to integrate the four fundamental concepts of feng shui knowledge. It is recognized that a concept hierarchy approach provides a powerful way to represent structural knowledge (Tam, 1993). Using a concept hierarchy, the relationship between each piece of data can be expressed and presented with abstract terms (Lee *et al*, 1997).

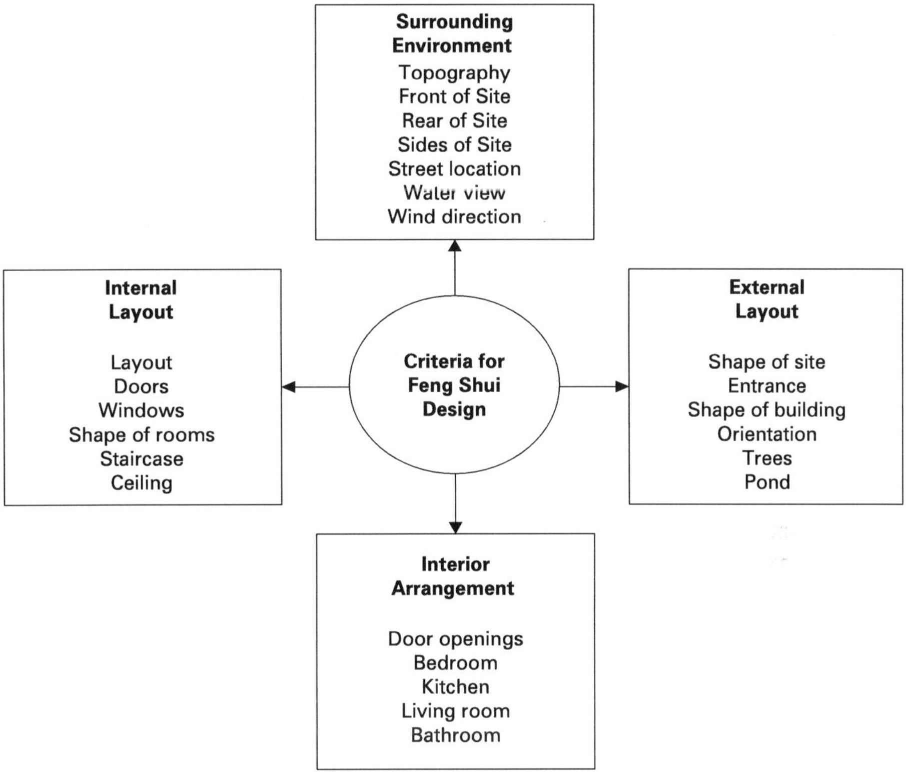

FIGURE 4 Feng shui criteria grouped in four design modules.
Source: Mak (2004)

A concept hierarchy takes the form of a generalization tree in which the most specific concepts (concept instances) form the leaves of the tree and the internal nodes represent intermediate concepts. Each intermediate concept is derived from generalizing concepts residing immediately at the next lower level. As a result, a path from the root to a leaf corresponds to a sequence of concepts, each obtained by adding more specific information to the previous one. Therefore, in a concept hierarchy, a concept needs only to store information specific to itself.

Construction of a concept hierarchy requires summarizing and generalizing a knowledge domain; however, it is still an unstructured process that involves a manual development of the hierarchy (Zupan *et al*, 1999). Concept hierarchies can be built explicitly by domain experts or knowledge engineers or extracted from a database using structural information and data distribution; these form the basic sources of the domain knowledge (Lee *et al*, 1998). Tam (1993) suggested that to construct a concept hierarchy from a set of concept instances, the order of the concept instances is the most important feature, especially when the number of concept instances is small. The concept hierarchy constructed from the first few concept instances will form the skeleton of the hierarchy that will be expanded by adding the remaining instances.

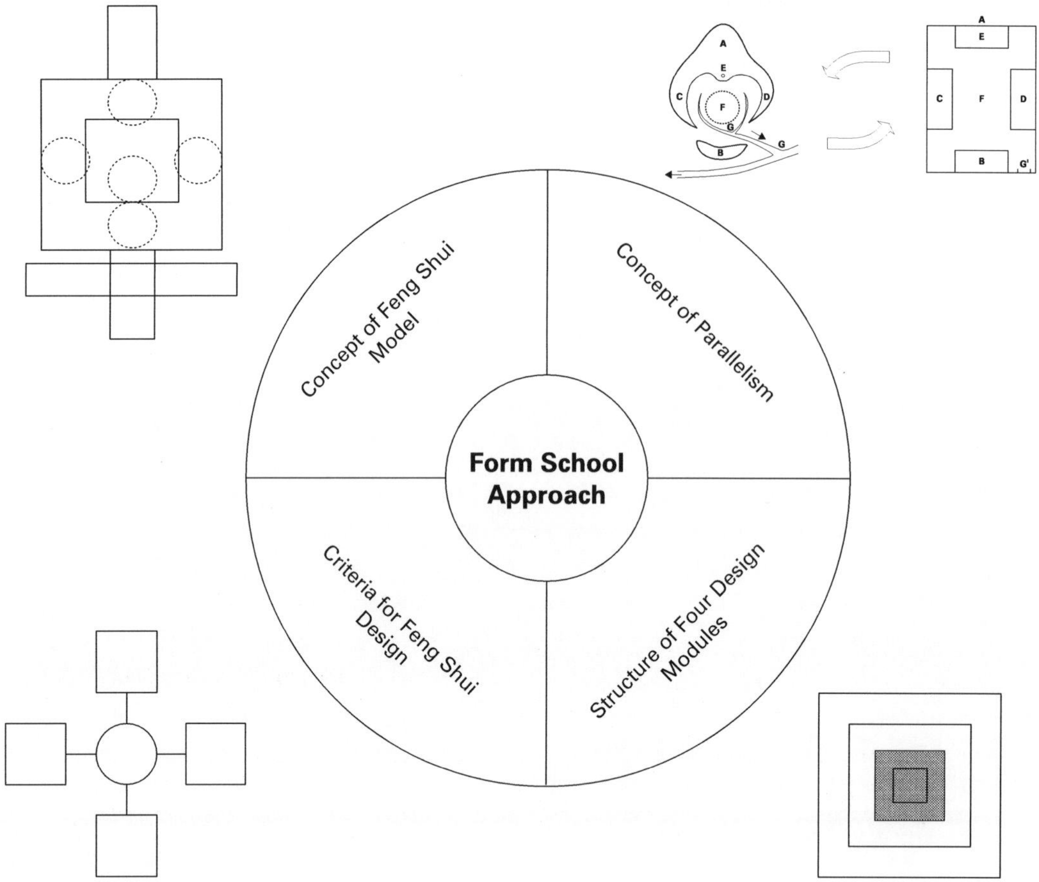

FIGURE 5 Organization of the Form School approach.
Source: Mak (2004)

HIERARCHIES OF FOUR FUNDAMENTAL CONCEPTS

When constructing a concept hierarchy of the feng shui knowledge structure, the four fundamental concepts of the Form School approach are first translated into sets of concept instances represented in a format of hierarchy to indicate the characteristics and structure within each concept:

- hierarchy of the feng shui model (Figure 6)
- hierarchy of the feng shui model for exterior and interior (Figure 7)
- hierarchy of the four design modules (Figure 8) and
- hierarchy of the feng shui design criteria (Figure 9).

SKELETAL HIERARCHICAL STRUCTURE

A skeletal hierarchical structure is constructed to demonstrate the interrelationships between the four fundamental concepts of the Form School approach. These four fundamental concepts are first represented in a format of hierarchy and are then integrated to form a skeletal structure of the hierarchy of the feng shui knowledge (Figure 10). Primarily, the hierarchy of the four design modules (Figure 8) outlines the skeletal

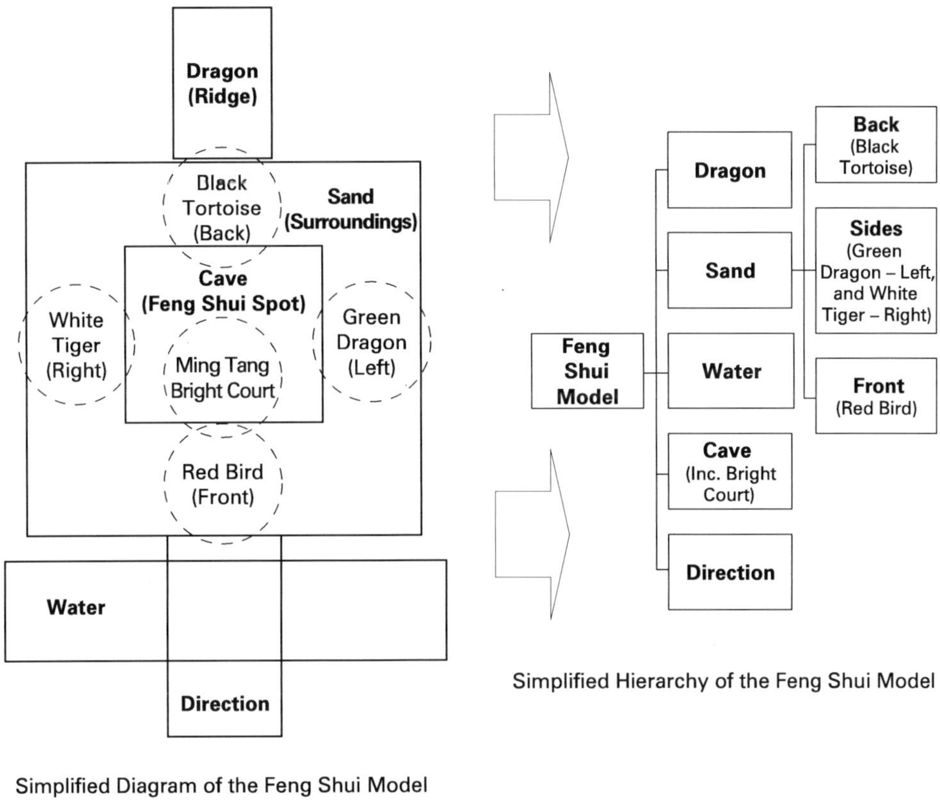

FIGURE 6 Hierarchy of the feng shui model.
Source: Mak (2004)

structure of the hierarchy. The hierarchy of feng shui knowledge is first classified into outer and inner forms, and then categorized into four design modules – the surrounding environment, the external layout, the internal layout and the interior arrangement. Each design module is connected to a feng shui model because the concept of parallelism implies that this feng shui model applies to the exterior as well as the interior (Figure 7). The concept of the feng shui model is incorporated with the hierarchy of feng shui design criteria which are grouped according to the four design modules (Figure 9).

After the skeletal hierarchical structure of the feng shui knowledge is created, the overall hierarchical structure of feng shui knowledge can be derived in detail to show further integration of the four fundamental concepts of the Form School approach (Mak, 2004; Mak *et al*, 2005). The overall hierarchy has more information and more elegant shape than that of the first hierarchy which only provides a skeletal hierarchy; the overall hierarchy tends to result in a more successful generalization process (Lee *et al*, 1998). For instance, in each design module, the hierarchy of the feng shui model can be organized in detail with the combination of the five geographical secrets and the four emblems theory as shown in Figure 6. The relationship between these attributes is then reviewed and connected to the relevant feng shui

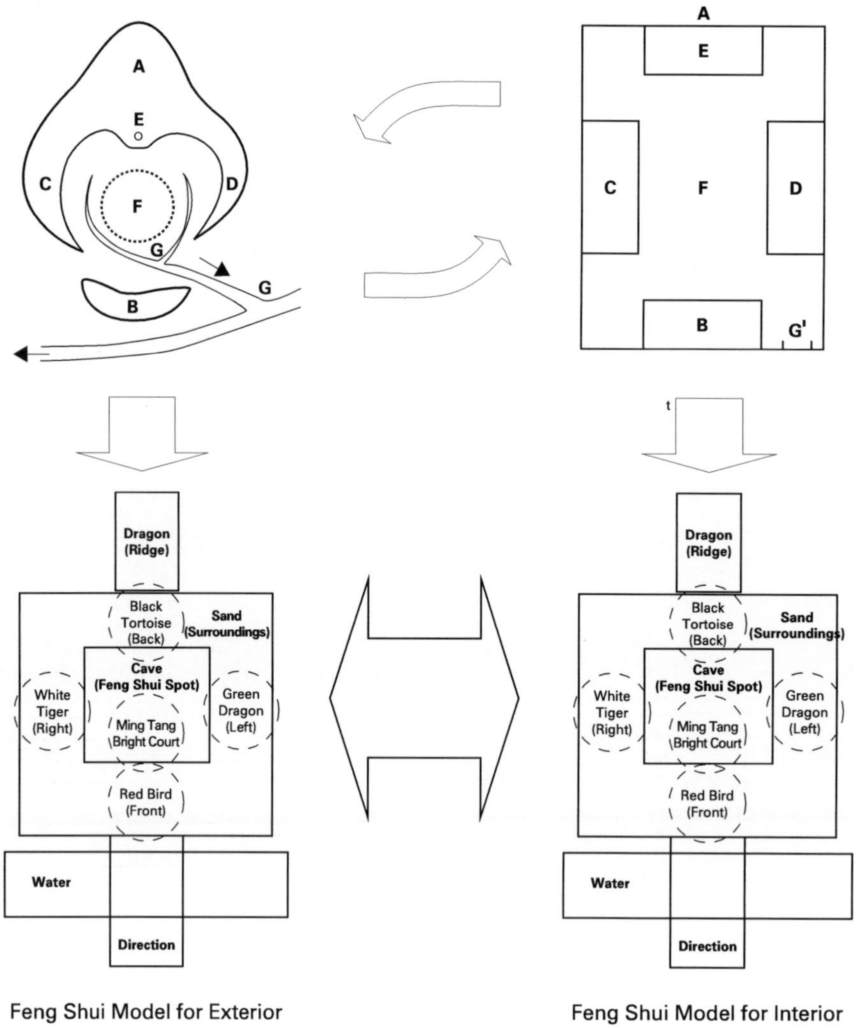

FIGURE 7 Hierarchy of the feng shui model for exterior and interior

criteria in the same design module to construct the overall hierarchical structure. This hierarchical structure will be further developed to incorporate specific design conditions related to each of the feng shui criteria.

A questionnaire survey of architects in Sydney and Hong Kong was used to evaluate the four fundamental concepts of the Form School approach and the developed hierarchical structure of the conceptual framework. The results of the survey demonstrated that these fundamental concepts and the conceptual framework are accepted by architects (Mak, 2004). Overall, there was very strong correlation between the groups of architects in Sydney and Hong Kong notwithstanding the fact that they have very different cultural, educational and geographical backgrounds (Mak and Ng, 2005).

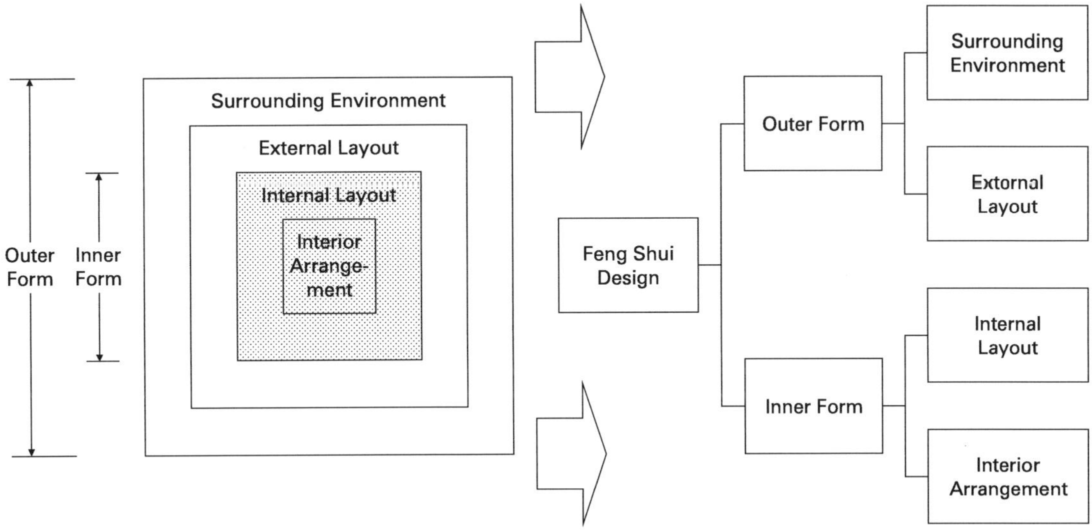

FIGURE 8 Hierarchy of the four design modules.
Source: Mak (2004)

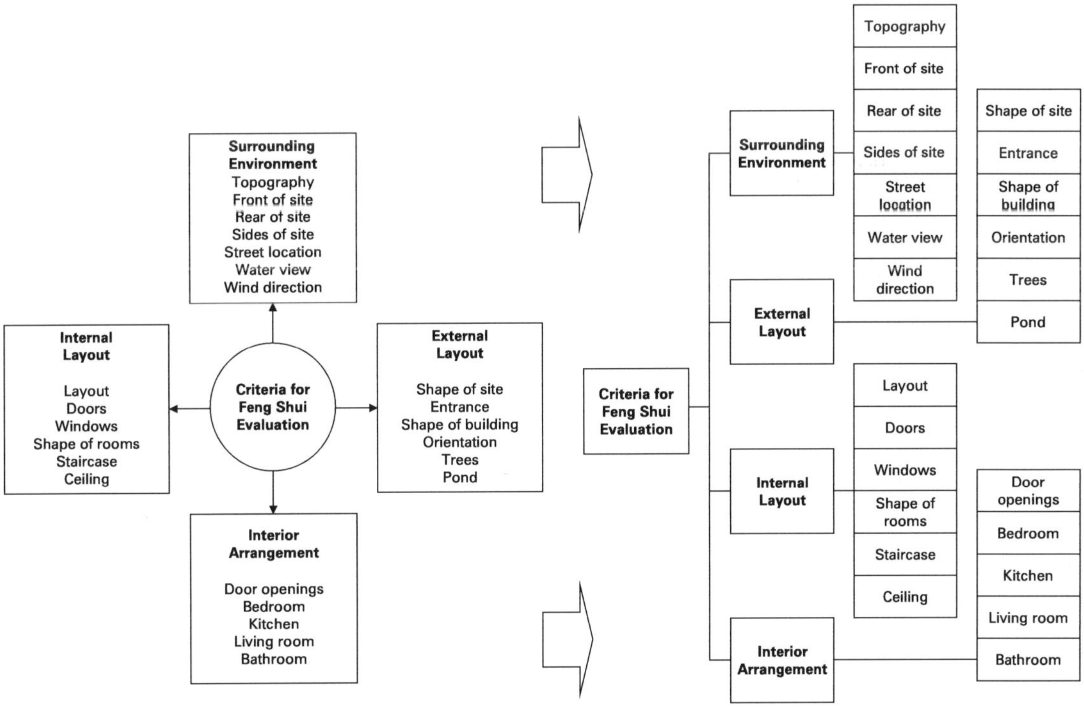

FIGURE 9 Hierarchy of the feng shui design criteria.
Source: Mak (2004)

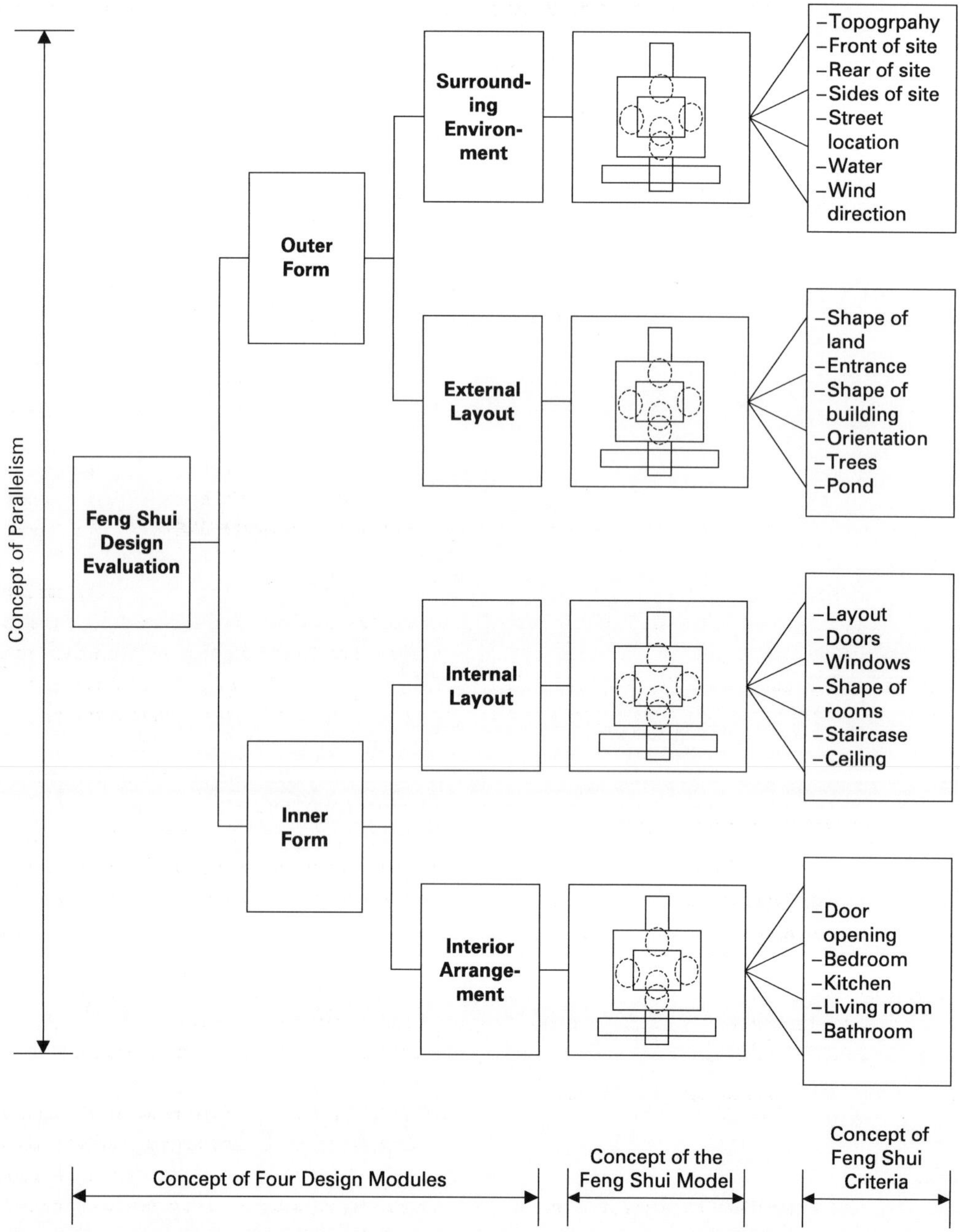

FIGURE 10 Skeletal hierarchical structure of feng shui knowledge.
Source: Mak *et al* (2005)

CONCLUSIONS

Building design is recognized as a very complex and intuitive process. Architects use their knowledge accumulated from intuition and past experiences to make heuristic design decisions. These characteristics are similar to the Chinese philosophy of direct insight into the nature and principles and practices of feng shui as applied in ancient Chinese architecture. It is suggested that interpreting feng shui knowledge would enable the development of an alternative design framework from this Chinese architectural discipline. In particular, the Form School provides a holistic approach that allows integrated components and elements to be considered for the built environment. However, due to a lack of properly documented feng shui principles and guidelines, it is difficult for architects to apply the knowledge of feng shui in further enhancing their design. There is a need to establish a more structured code of practice for feng shui application before the theories can be widely spread and adopted.

Although the principles and practices of the Form School approach provide a broad understanding of feng shui knowledge, a systematic development of the four fundamental concepts and a hierarchical structure of feng shui knowledge have never been established. In this paper, the authors are using the knowledge elicitation and representation techniques to establish a more structured framework of feng shui, and the results may help to improve the awareness and usage of this ancient wisdom for building design.

AUTHOR CONTACT DETAILS

Michael Mak (corresponding author): School of Architecture and Built Environment, University of Newcastle, NSW 2308, Australia. Tel: +612 4921 7450, fax: +612 4921 6913, e-mail: Michael.Mak@newcastle.edu.au

S. Thomas Ng: Department of Civil Engineering, University of Hong Kong, Pokfulam Road, Hong Kong.

REFERENCES

Archer, L.B., 1969, 'The structure of the design process', in G. Broadbent and T. Ward (eds), *Design Methods in Architecture*, London, Lund Humphries, 76–102.

Asimow, M., 1962, *Introduction to Design*, Englewood Cliffs, Prentice Hall.

Cheng, J. and Kong, S., 1993, *Feng Shui and Architecture*, Nanchang, Jiangxi Science and Technology Press (in Chinese).

Chiou, S.-C. and Krishnamurti, R., 1997, 'Unraveling feng shui', in *Environment and Planning B: Planning and Design*, 24(4), 549–572.

Choy, H., 1999, 'Go with the flow', in *Residential Property Investors Guide*, Sydney, Property Council of Australia, 19.

Feuchtwang, S.D.R., 1974, *An Anthropological Analysis of Chinese Geomancy*, Vientiane, Laos, Vithagna.

Glaser, B.G. and Strauss, A.L., 1967, *The Discovery of Grounded Theory: Strategies for Qualitative Research*, New York, Aldine De Gruyter.

Gregory, S.A. (ed), 1966, *The Design Method*, London, Butterworth.

Han, K.-T., 1995, *Basic Theory of Landscape Feng Shui*, Taipei, Lamper Enterprises (in Chinese).

He, X., 1990, *The Source of Feng Shui*, Nanjing, Southeast University Press (in Chinese).

He, X., 1998, *Feng Shui: Chinese Tradition in a Manchester Context*, PhD thesis, University of Manchester.

He, X. and Luo, J., 1995, *History of Chinese Feng Shui*, Shanghai, Shanghai Arts and Literature Press (in Chinese).

Humphreys, C., 1976, *Zen Buddhism*, London, Allen and Unwin.

Hwangbo, A.B., 1999, *In Search of Alternative Traditions in Architecture: a Cross-Cultural Interdisciplinary Study*, PhD thesis in architecture, University of Sheffield.

Jones, J.C., 1970, *Design Methods*, New York, John Wiley and Sons.

Jones, J.C., 1963, 'A method of systematic design', in J.C. Jones and D. Thronley (eds), *Conference on Design Methods*, Oxford, Pergamon.

Jung, I., 1996, *A Knowledge-based Approach to Computer-aided Architectural Design and Evaluation using Cost Estimation as an Illustration*, PhD thesis, University of Sheffield.

Kuhn, T.S., 1970, *The Structure of Scientific Revolutions*, 2nd edn, Chicago, University of Chicago Press.

Lee, D.H., Kim, S.W., Methou, G. and Lee, J.Y., 1997, 'Search for the blind: a schema independent query method using concept hierarchy and keywords', in *Computers and Industrial Engineering*, 33(3–4), 657–660.

Lee, D.H., Kim, S.M., Nam, D.W. and Lee, J.Y., 1998, 'A concept hierarchy management system for knowledge discovery in database', in *Proceedings of the Joint Conference on Information Sciences*, Durham, NC, Duke University Press, 3, 298–301.

Lee, S.-H., 1986, *Feng Shui: Its Context and Meaning*, PhD thesis, Cornell University.

Lip, E., 1979, *Chinese Geomancy*, Singapore, Times Books International.

Lip, E., 1986, *Feng Shui for the Home*, Singapore, Heian International.

Mak, M.Y., 2004, *Application of Feng Shui Knowledge to Preliminary Design Evaluation using Knowledge-based Expert Systems Approach*, PhD thesis, University of Newcastle, Australia.

Mak, M.Y. and Ng, S.T., 2005, 'The art and science of feng shui – a study on architects' perception', in *Building and Environment*, 40(3), 427–434.

Mak, M.Y., Ng, S.T. and Chen, S.E., 2005, 'Formulating a hierarchical structure of feng shui knowledge', in F. Loforte *et al* (eds), *Information and Knowledge Management in a Global Economy*, Lisbon, Portugal, Instituto Superior Técnico, 395–404.

Poincare, H., 1929, *The Foundations of Science*, New York, Science House Inc.

Popper, K.R., 1972, *Conjectures and Refutations: The Growth of Scientific Knowledge*, 4th edn, London, Routledge and Kegan Paul.

Raftery, J., 1991, *Principles of Building Economics*, London, Blackwell Scientific Publications.

Rossbach, S., 1984, *Feng Shui*, London, Rider.

Rossbach, S., 1987, *Interior Design with Feng Shui*, London, Rider.

Sargent, P., 1994, 'Design science or nonscience', in *Design Studies*, 15(4), 389–402.

Shang, K., 1992, 'China's pattern of feng shui: its formation, relationship to environment and landscaping', in Q. Wang (ed), *Research of Feng Shui Theory*, Tianjin, Tianjin University Press, 26–32 (in Chinese).

Simon, H.A., 1969, *The Sciences of the Artificial*, Cambridge, The MIT Press.

Skinner, S., 1982, *The Living Earth Manual of Feng Shui Chinese Geomancy*, London, Arkana.

Suckle, A., 1980, *By Their Own Design*, New York, Granada.

Tam, K.Y., 1993, 'Applying conceptual clustering to knowledge-based construction', in *Decision Support Systems*, 10(2), 173–198.

Too, L., 1996, *The Complete Illustrated Guide to Feng Shui*, London, Element.

Wang, Q. (ed), 1992, *Research of Feng Shui Theory*, Tianjin, Tianjin University Press (in Chinese).

Xu, P., 1990, *Feng Shui: A Model for Landscape Analysis*, DrDes thesis, Harvard University.

Yi, D., Yu, L. and Hong, Y., 1996, *Geomancy and the Selection of Architecture Placement in Ancient China*, Shi Jia Zhuang, Hebei Science and Technology Press (in Chinese).

Zupan, B., Bohanec, M., Demsar, J. and Bratko, I., 1999, 'Learning by discovering concept hierarchies', in *Artificial Intelligence*, 109(1–2), 211–242.